Julian Ralph, Frederic Remington

On Canada's Frontier

Sketches of history, sport, and adventure, and of the Indians, missionaries, fur-traders, and newer settlers of western Canada

Julian Ralph, Frederic Remington

On Canada's Frontier

Sketches of history, sport, and adventure, and of the Indians, missionaries, fur-traders, and newer settlers of western Canada

ISBN/EAN: 9783337148928

Printed in Europe, USA, Canada, Australia, Japan

Cover: Foto ©ninafisch / pixelio.de

More available books at **www.hansebooks.com**

ON CANADA'S FRONTIER

Sketches

OF HISTORY, SPORT, AND ADVENTURE
AND OF THE INDIANS, MISSIONARIES
FUR-TRADERS, AND NEWER SETTLERS
OF WESTERN CANADA

BY

JULIAN RALPH

ILLUSTRATED

NEW YORK
HARPER & BROTHERS, FRANKLIN SQUARE
1892

Copyright, 1892, by HARPER & BROTHERS.

All rights reserved.

TO
THE PEOPLE OF CANADA

THIS BOOK IS GRATEFULLY DEDICATED BY THE AUTHOR
WHO, DURING MANY LONG JOURNEYS IN THE CANADIAN WEST
WAS ALWAYS AND EVERYWHERE TREATED WITH AN EXTREME
FRIENDLINESS TO WHICH HE HERE TESTIFIES
BUT WHICH HE CANNOT EASILY RETURN
IN EQUAL MEASURE

PREFACE

If all those into whose hands this book may fall were as well informed upon the Dominion of Canada as are the people of the United States, there would not be needed a word of explanation of the title of this volume. Yet to those who might otherwise infer that what is here related applies equally to all parts of Canada, it is necessary to explain that the work deals solely with scenes and phases of life in the newer, and mainly the western, parts of that country. The great English colony which stirs the pages of more than two centuries of history has for its capitals such proud and notable cities as Montreal, Quebec, Toronto, Halifax, and many others, to distinguish the progressive civilization of the region east of Lake Huron—the older provinces. But the Canada of the geographies of to-day is a land of greater area than the United States; it is, in fact, the "British America" of old. A great trans-Canadian railway has joined the ambitious province of the Pacific slope to the provinces of old Canada with stitches of steel across the Plains. There the same mixed surplusage of Europe that settled our own West is elbowing the fur-trader and the Indian out of the way, and is laying out farms far north, in the smiling Peace River district, where it was only a little while ago supposed that there were but two seasons, winter and late spring. It is with that new part of Canada, between the ancient and well-populated provinces and the sturdy new cities of the Pacific Coast, that this book deals. Some references to the North are added in those chapters that treat of hunting and fishing and fur-trading.

The chapters that compose this book originally formed a series of papers which recorded journeys and studies made in Canada during the past three years. The first one to be published was that which describes a settler's colony in which a few titled foreigners took the lead; the others were written so recently that they should possess the same interest and value as if they here first met the public eye. What that interest and value amount to is for the reader to judge, the author's position being such that he may only call attention to the fact that he had access to private papers and documents when he prepared the sketches of the Hudson Bay Company, and that, in pursuing information about the great province of British Columbia, he was not able to learn that a serious and extended study of its resources had ever been made. The principal studies and sketches were prepared for and published in HARPER'S MAGAZINE. The spirit in which they were written was solely that of one who loves the open air and his fellow-men of every condition and color, and who has had the good-fortune to witness in newer Canada something of the old and almost departed life of the plainsmen and woodsmen, and of the newer forces of nation-building on our continent.

CONTENTS

		PAGE
I.	TITLED PIONEERS	1
II.	CHARTERING A NATION	11
III.	A FAMOUS MISSIONARY	53
IV.	ANTOINE'S MOOSE-YARD	66
V.	BIG FISHING	115
VI.	"A SKIN FOR A SKIN"	134
VII.	"TALKING MUSQUASH"	190
VIII.	CANADA'S EL DORADO	244
IX.	DAN DUNN'S OUTFIT	290

ILLUSTRATIONS

	PAGE
The Romantic Adventure of Old Sun's Wife	Frontispiece
Dr. Rudolph Meyer's Place on the Pipestone	2
Settler's Sod Cabin	3
Whitewood, a Settlement on the Prairie	4
Interior of Sod Cabin on the Frontier	5
Prairie Sod Stable	7
Trained Ox Team	9
Indian Boys Running a Foot-race	31
Indian Mother and Boy	36
Opening of the Soldier Clan Dance	39
Sketch in the Soldier Clan Dance	43
A Fantasy from the Pony War-dance	47
Throwing the Snow Snake	51
Father Lacombe Heading the Indians	61
The Hotel—Last Sign of Civilization	69
"Give me a light"	73
Antoine, from Life	79
The Portage Sleigh on a Lumber Road	83
The Track in the Winter Forest	87
Pierre, from Life	91
Antoine's Cabin	93
The Camp at Night	97
A Moose Bull Fight	101
On the Moose Trail	103
In Sight of the Game—"Now Shoot"	105
Success	109
Hunting the Caribou—"Shoot! Shoot!"	111
Indians Hauling Nets on Lake Nepigon	119
Trout-fishing Through the Ice	127
Rival Traders Racing to the Indian Camp	137
The Bear-trap	143

ILLUSTRATIONS

	PAGE
Huskie Dogs Fighting	147
Painting the Robe	151
Coureur du Bois	159
A Fur-trader in the Council Tepee	163
Buffalo Meat for the Post	167
The Indian Hunter of 1750	171
Indian Hunter Hanging Deer Out of the Reach of Wolves	173
Making the Snow-shoe	177
A Hudson Bay Man (Quarter-breed)	181
The Coureur du Bois and the Savage	185
Talking Musquash	193
Indian Hunters Moving Camp	198
Setting a Mink-trap	201
Wood Indians Come to Trade	205
A Voyageur, or Canoe-man, of Great Slave Lake	209
In a Stiff Current	211
Voyageur with Tumpline	217
Voyageurs in Camp for the Night	221
"Huskie" Dogs on the Frozen Highway	227
The Factor's Fancy Toboggan	233
Halt of a York Boat Brigade for the Night	239
An Impression of Shuswap Lake, British Columbia	251
The Tschummum, or Tool Used in Making Canoes	257
The First of the Salmon Run, Fraser River	261
Indian Salmon-fishing in the Thrasher	266
Going to the Potlatch—Big Canoe, North-west Coast	269
The Salmon Cache	275
An Ideal of the Coast	279
The Potlatch	283
An Indian Canoe on the Columbia	293
"You're setting your nerves to stand it"	297
Jack Kirkup, the Mountain Sheriff	299
Engineer on the Preliminary Survey	303
Falling Monarchs	308
Dan Dunn on His Works	311
The Supply Train Over the Mountain	313
A Sketch on the Work	317
The Mess Tent at Night	319
"They Gained Erectness by Slow Jolts"	322

ON CANADA'S FRONTIER

I

TITLED PIONEERS

THERE is a very remarkable bit of this continent just north of our State of North Dakota, in what the Canadians call Assiniboia, one of the North-west Provinces. Here the plains reach away in an almost level, unbroken, brown ocean of grass. Here are some wonderful and some very peculiar phases of immigration and of human endeavor. Here is Major Bell's farm of nearly one hundred square miles, famous as the Bell Farm. Here Lady Cathcart, of England, has mercifully established a colony of crofters, rescued from poverty and oppression. Here Count Esterhazy has been experimenting with a large number of Hungarians, who form a colony which would do better if those foreigners were not all together, with only each other to imitate—and to commiserate. But, stranger than all these, here is a little band of distinguished Europeans, partly noble and partly scholarly, gathered together in as lonely a spot as can be found short of the Rockies or the far northern regions of this continent.

DR. RUDOLPH MEYER'S PLACE ON THE PIPESTONE.

These gentlemen are Dr. Rudolph Meyer, of Berlin, the Comte de Cazes and the Comte de Raffignac, of France, and M. le Bidau de St. Mars, of that country also. They form, in all probability, the most distinguished and aristocratic little band of immigrants and farmers in the New World.

Seventeen hundred miles west of Montreal, in a vast prairie where settlers every year go mad from loneliness, these polished Europeans till the soil, strive for prizes at the provincial fairs, fish, hunt, read the current literature of two continents, and are happy. The soil in that region is of remarkable depth and richness, and is so black that the roads and cattle-trails look like ink lines on brown paper. It is part of a vast territory of uniform appearance, in one portion of which are the richest wheat-lands of the continent. The Canadian Pacific Railway crosses Assiniboia, with stops about five miles apart —some mere stations and some small settlements. Here the best houses are little frame dwellings; but very many of the settlers live in shanties made of sods, with such thick walls and tight roofs, all of sod, that the awful winters, when the mercury falls to forty degrees below zero, are endured in them better than in the more costly frame dwellings.

I stopped off the cars at Whitewood, picking that four-year-old village out at hap-hazard as a likely point at which to see how the immigrants live in a brand-new country. I had no idea of the existence of any of the persons I found there. The most perfect hospitality is offered to strangers in such infant communities, and while enjoying the shelter of a merchant's house I obtained news of the distinguished

SETTLER'S SOD CABIN

settlers, all of whom live away from the railroad in solitude not to be conceived by those who think their homes the most isolated in the older parts of the country. I had only time to visit Dr. Rudolph Meyer, five miles from Whitewood, in the valley of the Pipestone.

WHITEWOOD, A SETTLEMENT ON THE PRAIRIE

The way was across a level prairie, with here and there a bunch of young wolf-willows to break the monotonous scene, with tens of thousands of gophers sitting boldly on their haunches within reach of the wagon whip, with a sod house in sight in one direction at one time and a frame house in view at another. The talk of the driver was spiced with news of abundant wild-fowl, fewer deer, and marvellously numerous small quadrupeds, from wolves and foxes down. He talked of bachelors living here and there alone on that sea of grass, for all the world like men in small boats on the ocean; and I saw, contrariwise, a man and wife who blessed Heaven for an unheard-of number of children, especially prized because each new-comer lessened the loneliness. I heard of the long and dreadful winters when the snowfall is so light that horses and mules may always paw down to grass, though cattle stand and starve and freeze to death. I heard, too, of the way the snow comes in flurried squalls, in which men are lost within pistol-shot of their homes. In time the wagon came to

INTERIOR OF SOD CABIN ON THE FRONTIER.

a sort of coulee or hollow, in which some mechanics imported from Paris were putting up a fine cottage for the Comte de Raffignac. Ten paces farther, and I stood on the edge of the valley of the Pipestone, looking at a scene so poetic, pastoral, and beautiful that in the whole transcontinental journey there were few views to compare with it.

Reaching away far below the level of the prairie was a bowl-like valley, a mile long and half as wide, with a crystal stream lying like a ribbon of silver midway between its sloping walls. Another valley, longer yet, served as an extension to this. On the one side the high grassy walls were broken with frequent gullies, while on the other side was a park-like growth of forest trees. Meadows and fields lay between, and nestling against the eastern or grassy wall was the quaint, old-fashioned German house of the learned doctor. Its windows looked out on those beautiful little valleys, the property of the doctor—a little world far below the great prairie out of which sportive and patient Time had hollowed it. Externally the long, low, steep-roofed house was German, ancient, and picturesque in appearance. Its main floor was all enclosed in the sash and glass frame of a covered porch, and outside of the walls of glass were heavy curtains of straw, to keep out the sun in summer and the cold in winter. In-doors the house is as comfortable as any in the world. Its framework is filled with brick, and its trimmings are all of pine, oiled and varnished. In the heart of the house is a great Russian stove—a huge box of brickwork, which is filled full of wood to make a fire that

PRAIRIE SOD STABLE

is made fresh every day, and that heats the house for twenty-four hours. A well-filled wine-cellar, a well-equipped library, where HARPER'S WEEKLY, and *Über Land und Mer*, *Punch*, *Puck*, and *Die Fliegende Blätter* lie side by side, a kindly wife, and a stumbling baby, tell of a combination of domestic joys that no man is too rich to envy. The library is the doctor's workshop He is now engaged in compiling a digest of the economic laws of nations. He is already well known as the author of a *History of Socialism* (in Germany, the United States, Scandinavia, Russia, France, Belgium, and elsewhere), and also for his *History of Socialism in Germany*. He writes in French and German, and his works are published in Germany.

Dr. Meyer is fifty-three years old. He is a political exile, having been forced from Prussia for connection with an unsuccessful opposition to Bismarck. It is because he is a scholar seeking rest from the turmoil of politics that one is able to comprehend his living in this overlooked corner of the world. Yet when that is understood, and one knows what an Arcadia his little valley is, and how complete are his comforts within-doors, the placidity with which he smokes his pipe, drinks his beer, and is waited upon by servants imported from Paris, becomes less a matter for wonder than for congratulation. He has shared part of one valley with the Comte de Raffignac, who thinks there is nothing to compare with it on earth. The count has had his house built near the abruptly-broken edge of the prairie, so that he may look down upon the calm and beautiful valley and enjoy it, as he could not had he built in the valley itself. He is a youth of very old French family, who loves hunting and horses. He was contemplating the raising of horses for a business when I was there. But the count mars the romance of his membership in this little band by going to Paris now and then, as a young man would be likely to.

Out-of-doors one saw what untold good it does to the present and future settlers to have such men among them. The hot-houses, glazed vegetable beds, the plots of cultivated ground, the nurseries of young trees—all show at what cost of money and patience the Herr Doctor is experimenting with every tree and flower and vegetable and cereal to discover what can be grown with profit in that region of rich soil and short summers, and what cannot. He is in communication with the seedsmen, to say nothing of the savants, of Europe and this country, and whatever he plants is of the best. Near his quaint dwelling he has a house for his gardener, a smithy, a tool-house, a barn, and a cheese-factory, for he makes gruyere cheese in great quantities. He also raises horses and cattle.

The Comte de Cazes has a sheltered, favored claim a few miles to the northward, near the Qu' Appele River. He lives in great comfort, and is so successful a farmer that he carries off nearly all the prizes for the province, especially those given for prime vegetables. He has his wife and daughter and one of his sons with him, and an abundance of means, as, indeed, these distinguished settlers all appear to have.

TRAINED OX TEAM

These men have that faculty, developed in all educated and thinking souls, which enables them to banish loneliness and entertain themselves. Still, though Dr. Meyer laughs at the idea of danger, it must have been a little disquieting to live as he does during the Riel rebellion, especially as an Indian reservation is close by, and wandering red men are seen every day upon the prairie. Indeed, the Government thought fit to send men of the North-west Mounted Police to visit the doctor twice a week as lately as a year after the close of the half-breed uprising.

CHARTERING A NATION

HOW it came about that we chartered the Blackfoot nation for two days had better not be told in straightforward fashion. There is more that is interesting in going around about the subject, just as in reality we did go around and about the neighborhood of the Indians before we determined to visit them.

In the first place, the most interesting Indian I ever saw—among many kinds and many thousands —was the late Chief Crowfoot, of the Blackfoot people. More like a king than a chief he looked, as he strode upon the plains, in a magnificent robe of white bead-work as rich as ermine, with a gorgeous pattern illuminating its edges, a glorious sun worked into the front of it, and many artistic and chromatic figures sewed in gaudy beads upon its back. He wore an old white chimney-pot hat, bound around with eagle feathers, a splendid pair of *chaperajos*, all worked with beads at the bottoms and fringed along the sides, and bead-worked moccasins, for which any lover of the Indian or collector of his paraphernalia would have exchanged a new Winchester rifle without a second's hesitation. But though Crowfoot was so royally clothed, it was in himself that the kingly quality was most apparent. His face was extraordinarily like what portraits we have of Julius Cæsar, with the difference that Crowfoot had the complexion of an Egyptian mummy. The high forehead, the great aquiline nose, the thin lips, usually closed, the small, round, protruding chin, the strong jawbones, and the keen gray eyes composed a face in which every feature was finely moulded, and in which the warrior, the commander, and the counsellor were strongly suggested. And in each of these roles he played the highest part among the Indians of Canada from the moment that the whites and the red men contested the dominion of the plains until he died, a short time ago.

He was born and lived a wild Indian, and though the good fathers of the nearest Roman Catholic mission believe that he died a Christian, I am constrained to see in the reason for their thinking so only another proof of the consummate shrewdness of Crowfoot's life-long policy. The old king lay on his death-bed in his great wig-a-wam, with twenty-seven of his medicine-men around him, and never once did he pretend that he despised or doubted their magic. When it was evident that he was about to die, the conjurers ceased their long-continued, exhausting formula of howling, drumming, and all the rest, and, Indian-like, left Death to take his own. Then it was that one of the watchful, zealous priests, whose lives have indeed been like those of fathers to the wild Indians, slipped into the great tepee and administered the last sacrament to the old pagan.

"Do you believe?" the priest inquired.

"Yes, I believe," old Crowfoot grunted. Then he whispered, "But don't tell my people."

Among the last words of great men, those of Saponaxitaw (his Indian name) may never be recorded,
but to the student of the American aborigine they
betray more that is characteristic of the habitual attitude of mind of the wild red man towards civilizing
influences than any words I ever knew one to utter.

As the old chief crushed the bunch-grass beneath
his gaudy moccasins at the time I saw him, and as
his lesser chiefs and headmen strode behind him, we
who looked on knew what a great part he was bearing and had taken in Canada. He had been chief of
the most powerful and savage tribe in the North, and
of several allied tribes as well, from the time when
the region west of the Mississippi was *terra incognita*
to all except a few fur traders and priests. His warriors ruled the Canadian wilderness, keeping the
Ojibbeways and Crees in the forests to the east and
north, routing the Crows, the Stonies, and the Big-
Bellies whenever they pleased, and yielding to no
tribe they met except the Sioux to the southward
in our territory. The first white man Crowfoot
ever knew intimately was Father Lacombe, the noble
old missionary, whose fame is now world-wide among
scholars. The peaceful priest and the warrior chief
became fast friends, and from the day when the white
men first broke down the border and swarmed upon
the plains, until at the last they ran what Crowfoot
called their "fire-wagons" (locomotives) through his
land, he followed the priest's counselling in most important matters. He treated with the authorities,
and thereafter hindered his braves from murder,
massacre, and warfare. Better than that, during the

Riel rebellion he more than any other man, or twenty men, kept the red man of the plains at peace when the French half-breeds, led by their mentally irresponsible disturber, rebelled against the Dominion authorities.

When Crowfoot talked, he made laws. While he spoke, his nation listened in silence. He had killed as many men as any Indian warrior alive; he was a mighty buffalo-slayer; he was torn, scarred, and mangled in skin, limb, and bone. He never would learn English or pretend to discard his religion. He was an Indian after the pattern of his ancestors. At eighty odd years of age there lived no red-skin who dared answer him back when he spoke his mind. But he was a shrewd man and an archdiplomatist. Because he had no quarrel with the whites, and because a grand old priest was his truest friend, he gave orders that his body should be buried in a coffin, Christian fashion, and as I rode over the plains in the summer of 1890 I saw his burial-place on top of a high hill, and knew that his bones were guarded night and day by watchers from among his people. Two or three days before he died his best horse was slaughtered for burial with him. He heard of it. "That was wrong," he said; "there was no sense in doing that; and besides, the horse was worth good money." But he was always at least as far as that in advance of his people, and it was natural that not only his horse, but his gun and blankets, his rich robes, and plenty of food to last him to the happy hunting-grounds, should have been buried with him.

There are different ways of judging which is the best Indian, but from the stand-point of him who would examine that distinct product of nature, the Indian as the white man found him, the Canadian Blackfeet are among if not quite the best. They are almost as primitive and natural as any, nearly the most prosperous, physically very fine, the most free from white men's vices. They are the most reasonable in their attitude towards the whites of any who hold to the true Indian philosophy. The sum of that philosophy is that civilization gets men a great many comforts, but bundles them up with so many rules and responsibilities and so much hard work that, after all, the wild Indian has the greatest amount of pleasure and the least share of care that men can hope for. That man is the fairest judge of the red-skins who considers them as children, governed mainly by emotion, and acting upon undisciplined impulse; and I know of no more hearty, natural children than the careless, improvident, impulsive boys and girls of from five to eighty years of age whom Crowfoot turned over to the care of Three Bulls, his brother.

The Blackfeet of Canada number about two thousand men, women, and children. They dwell upon a reserve of nearly five hundred square miles of plains land, watered by the beautiful Bow River, and almost within sight of the Rocky Mountains. It is in the province of Alberta, north of our Montana. There were three thousand and more of these Indians when the Canadian Pacific Railway was built across their hunting-ground, seven or eight years ago, but they

are losing numbers at the rate of two hundred and
fifty a year, roughly speaking. Their neighbors, the
tribes called the Bloods and the Piegans, are of the
same nation. The Sarcis, once a great tribe, became weakened by disease and war, and many years
ago begged to be taken into the confederation.
These tribes all have separate reserves near to one
another, but all have heretofore acknowledged each
Blackfoot chief as their supreme ruler. Their old
men can remember when they used to roam as far
south as Utah, and be gone twelve months on the
war-path and on their foraging excursions for horses.
They chased the Crees as far north as the Crees
would run, and that was close to the arctic circle.
They lived in their war-paint and by the chase. Now
they are caged. They live unnaturally and die as
unnaturally, precisely like other wild animals shut up
in our parks. Within their park each gets a pound
of meat with half a pound of flour every day. Not
much comes to them besides, except now and then a
little game, tobacco, and new blankets. They are so
poorly lodged and so scantily fed that they are not
fit to confront a Canadian winter, and lung troubles
prey among them.

It is a harsh way to put it (but it is true of our own
government also) to say that one who has looked the
subject over is apt to decide that the policy of the
Canadian Government has been to make treaties with
the dangerous tribes, and to let the peaceful ones
starve. The latter do not need to starve in Canada,
fortunately; they trust to the Hudson Bay Company
for food and care, and not in vain. Having treated

with the wilder Indians, the rest of the policy is to send the brightest of their boys to trade-schools, and to try to induce the men to till the soil. Those who do so are then treated more generously than the others. I have my own ideas with which to meet those who find nothing admirable in any except a dead Indian, and with which to discuss the treatment and policy the live Indian endures, but this is not the place for the discussion. Suffice it that it is not to be denied that between one hundred and fifty and two hundred Blackfeet are learning to maintain several plots of farming land planted with oats and potatoes. This they are doing with success, and with the further result of setting a good example to the rest. But most of the bucks are either sullenly or stupidly clinging to the shadow and the memory of the life that is gone.

It was a recollection of that life which they portrayed for us. And they did so with a fervor, an abundance of detail and memento, and with a splendor few men have seen equalled in recent years—or ever may hope to witness again.

We left the cars at Gleichen, a little border town which depends almost wholly upon the Blackfeet and their visitors for its maintenance. It has two stores —one where the Indians get credit and high prices (and at which the red men deal), and one at which they may buy at low rates for cash, wherefore they seldom go there. It has two hotels and a half-dozen railway men's dwellings, and, finally, it boasts a tiny little station or barracks of the North-west Mounted Police, wherein the lower of the two rooms is fitted with a desk, and hung with pistols, guns, handcuffs, and cartridge belts, while the upper room contains the cots for the men at night.

We went to the store that the Indians favor—just such a store as you see at any cross-roads you drive past in a summer's outing in the country—and there were half a dozen Indians beautifying the door-way and the interior, like magnified majolica-ware in a crockery-shop. They were standing or sitting about with thoughtful expressions, as Indians always do when they go shopping; for your true Indian generates such a contemplative mood when he is about to spend a quarter that one would fancy he must be the most prudent and deliberate of men, instead of what he really is—the greatest prodigal alive except the negro. These bucks might easily have been mistaken for waxworks. Unnaturally erect, with arms folded beneath their blankets, they stood or sat without moving a limb or muscle. Only when a new-comer entered did they stir. Then they turned their heads deliberately and looked at the visitor fixedly, as eagles look at you from out their cages. They were strapping fine fellows, each bundled up in a colored blanket, flapping cloth leg-gear, and yellow moccasins. Each had the front locks of his hair tied in an upright bunch, like a natural plume, and several wore little brass rings, like baby finger-rings, around certain side locks down beside their ears.

There they stood, motionless and speechless, waiting until the impulse should move them to buy what they wanted, with the same deliberation with which they had waited for the original impulse which sent them to the store. If Mr. Frenchman, who kept the store, had come from behind his counter, English fashion, and had said: "Come, come; what d'you want? Speak up now, and be quick about it. No lounging here. Buy or get out." If he had said that, or anything like it, those Indians would have stalked out of his place, not to enter it again for a very long time, if ever. Bartering is a serious and complex performance to an Indian, and you might as well try to hurry an elephant up a gang-plank as try to quicken an Indian's procedure in trading.

We purchased of the Frenchman a chest of tea, a great bag of lump sugar, and a small case of plug tobacco for gifts to the chief. Then we hired a buckboard wagon, and made ready for the journey to the reserve.

The road to the reserve lay several miles over the plains, and commanded a view of rolling grass land, like a brown sea whose waves were petrified, with here and there a group of sickly wind-blown trees to break the resemblance. The road was a mere wagon track and horse-trail through the grass, but it was criss-crossed with the once deep ruts that had been worn by countless herds of buffalo seeking water.

Presently, as we journeyed, a little line of sand-hills came into view. They formed the Blackfoot cemetery. We saw the "tepees of the dead" here and there on the knolls, some new and perfect, some old and weather-stained, some showing mere tatters of cotton flapping on the poles, and still others only skeleton tents, the poles remaining and the cotton covering gone completely. We knew what we would see if we looked into those "dead tepees" (being careful to approach from the windward side). We would see, lying on the ground or raised upon a framework, a bundle that would be narrow at top and bottom, and broad in the middle—an Indian's body rolled up in a sheet of cotton, with his best bead-work and blanket and gun in the bundle, and near by a kettle and some dried meat and corn-meal against his feeling hungry on his long journey to the hereafter. As one or two of the tepees were new, we expected to see some family in mourning; and, sure enough, when we reached the great sheer-sided gutter which the Bow River has dug for its course through the plains, we halted our horse and looked down upon a lonely trio of tepees, with children playing around them and women squatted by the entrances. Three families had lost members, and were sequestered there in abject surrender to grief.

Those tents of the mourners were at our feet as we rode southward, down in the river gully, where the grass was green and the trees were leafy and thriving; but when we turned our faces to the eastward, where the river bent around a great promontory, what a sight met our gaze! There stood a city of tepees, hundreds of them, showing white and yellow and brown and red against the clear blue sky. A silent and lifeless city it seemed, for we were too far off to see the people or to hear their noises. The great huddle of little pyramids rose abruptly from the level bare grass against the flawless sky, not like one of those melancholy new treeless towns that white men are building all over the prairie, but rather like

a mosquito fleet becalmed at sea. There are two
camps on the Blackfoot Reserve, the North Camp
and the South Camp, and this town of tents was between the two, and was composed of more households
than both together; for this was the assembling for
the sun-dance, their greatest religious festival, and
hither had come Bloods, Piegans, and Sarcis as well
as Blackfeet. Only the mourners kept away; for
here were to be echoed the greatest ceremonials of
that dead past, wherein lives dedicated to war and to
the chase inspired the deeds of valor which each
would now celebrate anew in speech or song. This
was to be the anniversary of the festival at which the
young men fastened themselves by a strip of flesh in
their chests to a sort of Maypole rope, and tore their
flesh apart to demonstrate their fitness to be considered braves. At this feast husbands had the right
to confess their women, and to cut their noses off if
they had been untrue, and if they yet preferred life to
the death they richly merited. At this gala-time
sacrifices of fingers were made by brave men to the
sun. Then every warrior boasted of his prowess, and
the young beaus feasted their eyes on gayly-clad
maidens the while they calculated for what number
of horses they could be purchased of their parents.
And at each recurrence of this wonderful holiday-
time every night was spent in feasting, gorging, and
gambling. In short, it was the great event of the
Indian year, and so it remains. Even now you may
see the young braves undergo the torture; and if you
may not see the faithless wives disciplined, you may
at least perceive a score who have been, as well as

hear the mighty boasting, and witness the dancing, gaming, and carousing.

We turned our backs towards the tented field, for we had not yet introduced ourselves to Mr. Magnus Begg, the Indian agent in charge of the reserve. We were soon within his official enclosure, where a pretty frame house, an office no bigger than a freight car, and a roomy barn and stable were all overtopped by a central flag-staff, and shaded by flourishing trees. Mr. Begg was at home, and, with his accomplished wife, welcomed us in such a hearty manner as one could hardly have expected, even where white folks were so "mighty unsartin" to appear as they are on the plains. The agent's house without is like any pretty village home in the East; and within, the only distinctive features are a number of ornamental mounted wild-beast's heads and a room whose walls are lined about with rare and beautiful Blackfoot curios in skin and stone and bead-work. But, to our joy, we found seated in that room the famous chief Old Sun. He is the husband of the most remarkable Indian squaw in America, and he would have been Crowfoot's successor were it not that he was eighty-seven years of age when the Blackfoot Cæsar died. As chief of the North Blackfeet, Old Sun boasts the largest personal following on the Canadian plains, having earned his popularity by his fighting record, his commanding manner, his eloquence, and by that generosity which leads him to give away his rations and his presents. No man north of Mexico can dress more gorgeously than he upon occasion, for he still owns a buckskin outfit beaded to the value of a Worth gown. Moreover, he owns a red coat, such as the Government used to give only to great chiefs. The old fellow had lost his vigor when we saw him, and as he sat wrapped in his blanket he looked like a half-emptied meal bag flung on a chair. He despises English, but in that marvellous Volapük of the plains called the sign language he told us that his teeth were gone, his hearing was bad, his eyes were weak, and his flesh was spare. He told his age also, and much else besides, and there is no one who reads this but could have readily understood his every statement and sentiment, conveyed solely by means of his hands and fingers. I noticed that he looked like an old woman, and it is a fact that old Indian men frequently look so. Yet no one ever saw a young brave whose face suggested a woman's, though their beardless countenances and long hair might easily create that appearance.

Mr. Remington was anxious to paint Old Sun and his squaw, particularly the latter, and he easily obtained permission, although when the time for the mysterious ordeal arrived next day the old chief was greatly troubled in his superstitious old brain lest some mischief would befall him through the medium of the painting. To the Indian mind the sun, which they worship, has magical, even devilish, powers, and Old Sun developed a fear that the orb of day might "work on his picture" and cause him to die. Fortunately I found in Mr. L'Hereux, the interpreter, a person who had undergone the process without dire consequences, was willing to undergo it again, and who added that his father and mother had submitted to the operation, and yet had lived to a yellow old age. When Old Sun brought his wife to sit for her portrait I put all etiquette to shame in staring at her, as you will all the more readily believe when you know something of her history.

Old Sun's wife sits in the council of her nation—the only woman, white, red, or black, of whom I have ever heard who enjoys such a prerogative on this continent. She earned her peculiar privileges, if any one ever earned anything. Forty or more years ago she was a Piegan maiden known only in her tribe, and there for nothing more than her good origin, her comeliness, and her consequent value in horses. She met with outrageous fortune, but she turned it to such good account that she was speedily ennobled. She was at home in a little camp on the plains one day, and had wandered away from the tents, when she was kidnapped. It was in this wise: other camps were scattered near there. On the night before the day of her adventure a band of Crows stole a number of horses from a camp of the Gros Ventres, and very artfully trailed their plunder towards and close to the Piegan camp before they turned and made their way to their own lodges. When the Gros Ventres discovered their loss, and followed the trail that seemed to lead to the Piegan camp, the girl and her father, an aged chief, were at a distance from their tepees, unarmed and unsuspecting. Down swooped the Gros Ventres. They killed and scalped the old man, and then their chief swung the young girl upon his horse behind him, and binding her to him with thongs of buckskin, dashed off triumphantly

for his own village. That has happened to many
another Indian maiden, most of whom have behaved
as would a plaster image, saving a few days of weeping.
Not such was Old Sun's wife. When she and
her captor were in sight of the Gros Ventre village,
she reached forward and stole the chief's scalping-
knife out of its sheath at his side. With it, still wet
with her father's blood, she cut him in the back
through to the heart. Then she freed his body from
hers, and tossed him from the horse's back. Leaping
to the ground beside his body, she not only scalped
him, but cut off his right arm and picked up his gun,
and rode madly back to her people, chased most of
the way, but bringing safely with her the three greatest
trophies a warrior can wrest from a vanquished
enemy. Two of them would have distinguished any
brave, but this mere village maiden came with all
three. From that day she has boasted the right to
wear three eagle feathers.

Old Sun was a young man then, and when he
heard of this feat he came and hitched the requisite
number of horses to her mother's travois poles
beside her tent. I do not recall how many steeds
she was valued at, but I have heard of very high-
priced Indian girls who had nothing except their
feminine qualities to recommend them. In one case
I knew that a young man, who had been casting
what are called "sheep's eyes" at a maiden, went one
day and tied four horses to her father's tent. Then
he stood around and waited, but there was no sign
from the tent. Next day he took four more, and so
he went on until he had tied sixteen horses to the

tepee. At the least they were worth $20, perhaps $30, apiece. At that the maiden and her people came out, and received the young man so graciously that he knew he was "the young woman's choice," as we say in civilized circles, sometimes under very similar circumstances.

At all events, Old Sun was rich and powerful, and easily got the savage heroine for his wife. She was admitted to the Blackfoot council without a protest, and has since proven that her valor was not sporadic, for she has taken the war-path upon occasion, and other scalps have gone to her credit.

After a while we drove over to where the field lay littered with tepees. There seemed to be no order in the arrangement of the tents as we looked at the scene from a distance. Gradually the symptoms of a great stir and activity were observable, and we saw men and horses running about at one side of the nomad settlement, as well as hundreds of human figures moving in the camp. Then a nearer view brought out the fact that the tepees, which were of many sizes, were apt to be white at the base, reddish half-way up, and dark brown at the top. The smoke of the fires within, and the rain and sun without, paint all the cotton or canvas tepees like that, and very pretty is the effect. When closer still, we saw that each tepee was capped with a rude crown formed of pole ends—the ends of the ribs of each structure; that some of the tents were gayly ornamented with great geometric patterns in red, black, and yellow around the bottoms; and that others bore upon their sides rude but highly colored figures of

animals—the clan sign of the family within. Against
very many of the frail dwellings leaned a travois, the
triangle of poles which forms the wagon of the Indians. There were three or four very large tents,
the headquarters of the chiefs of the soldier bands
and of the head chief of the nation; and there was
one spotless new tent, with a pretty border painted
around its base, and the figure of an animal on either
side. It was the new establishment of a bride and
groom. A hubbub filled the air as we drew still
nearer; not any noise occasioned by our approach,
but the ordinary uproar of the camp—the barking of
dogs, the shouts of frolicking children, the yells of
young men racing on horseback and of others driving in their ponies. When we drove between the
first two tents we saw that the camp had been systematically arranged in the form of a rude circle,
with the tents in bunches around a great central
space, as large as Madison Square if its corners were
rounded off.

We were ushered into the presence of Three Bulls,
in the biggest of all the tents. By common consent
he was presiding as chief and successor to Crowfoot,
pending the formal election, which was to take place
at the feast of the sun-dance. European royalty
could scarcely have managed to invest itself with
more dignity or access to its presence with more
formality than hedged about this blanketed king.
He had assembled his chiefs and headmen to greet
us, for we possessed the eminence of persons bearing
gifts. He was in mourning for Crowfoot, who was his
brother, and for a daughter besides, and the form of

expression he gave to his grief caused him to wear
nothing but a flannel shirt and a breech-cloth, in
which he sat with his big brown legs bare and
crossed beneath him. He is a powerful man, with
an uncommonly large head, and his facial features, all
generously moulded, indicate amiability, liberality,
and considerable intelligence. Of middle age, smooth-
skinned, and plump, there was little of the savage in
his looks beyond what came of his long black hair.
It was purposely wore unkempt and hanging in his
eyes, and two locks of it were bound with many
brass rings. When we came upon him our gifts had
already been received and distributed, mainly to
three or four relatives. But though the others sat
about portionless, all were alike stolid and statuesque,
and whatever feelings agitated their breasts, whether
of satisfaction or disappointment, were equally hidden by all.

When we entered the big tepee we saw twenty-
one men seated in a circle against the wall and facing
the open centre, where the ground was blackened by
the ashes of former fires. Three Bulls sat exactly
opposite the queer door, a horseshoe-shaped hole
reaching two feet above the ground, and extended
by the partly loosened lacing that held the edges of
the tent-covering together. Mr. L'Hereux, the interpreter, made a long speech in introducing each of
us. We stood in the middle of the ring, and the
chief punctuated the interpreter's remarks with that
queer Indian grunt which it has ever been the custom to spell "ugh," but which you may imitate exactly if you will try to say " Ha " through your nose

while your mouth is closed. As Mr. L'Hereux is a great talker, and is of a poetic nature, there is no telling what wild fancy of his active brain he invented concerning us, but he made a friendly talk, and that was what we wanted. As each speech closed, Three Bulls lurched forward just enough to make the putting out of his hand a gracious act, yet not enough to disturb his dignity. After each salutation he pointed out a seat for the one with whom he had shaken hands. He announced to the council in their language that we were good men, whereat the council uttered a single "Ha" through its twenty-one noses. If you had seen the rigid stateliness of Three Bulls, and had felt the frigid self-possession of the twenty-one ramrod-mannered under-chiefs, as well as the deference which was in the tones of the other white men in our company, you would comprehend that we were made to feel at once honored and subordinate. Altogether we made an odd picture: a circle of men seated tailor fashion, and my own and Mr. Remington's black shoes marring the gaudy ring of yellow moccasins in front of the savages, as they sat in their colored blankets and fringed and befeathered gear, each with the calf of one leg crossed before the shin of the other.

But L'Hereux's next act after introducing us was one that seemed to indicate perfect indifference to the feelings of this august body. No one but he, who had spent a quarter of a century with them in closest intimacy, could have acted as he proceeded to do. He cast his eyes on the ground, and saw the mounds of sugar, tobacco, and tea heaped before only a certain few Indians. "Now who has done dose t'ing?" he inquired. "Oh, dat vill nevaire do 'tall. You haf done dose t'ing, Mistaire Begg? No? Who den? Chief? Nevaire mind. I make him all rount again, vaire deeferent. You shall see somet'ing." With that, and yet without ceasing to talk for an instant, now in Indian and now in his English, he began to dump the tea back again into the chest, the sugar into the bag, and the plug tobacco in a heap by itself. Not an Indian moved a muscle—unless I was right in my suspicion that the corners of Three Bulls' mouth curved upward slightly, as if he were about to smile. "Vot kind of wa-a-y to do-o somet'ing is dat?" the interpreter continued, in his sing-song tone. "You moos' haf one maje-dome [major-domo] if you shall try satisfy dose Engine." He always called the Indians "dose Engine." "Dat chief gif all dose present to his broders und cousins, vhich are in his famille. Now you shall see me, vot I shall do." Taking his hat, he began filling it, now with sugar and now with tea, and emptying it before some six or seven chiefs. Finally, when a double share was left, he gave both bag and chest to Three Bulls, to whom he also gave all the tobacco. "Such tam-fool peezness," he went on, "I do not see in all my life. I make visitation to de t'ree soljier chief vhich shall make one grand darnce for dose gentlemen, und here is for dose soljier chief not anyt'ing 'tall, vhile everyt'ing was going to one lot of beggaire relation of T'ree Bull. Dat is what I call one tam-fool way to do somet'ing."

The redistribution accomplished, Three Bulls wore

INDIAN BOYS RUNNING A FOOT-RACE

a grin of satisfaction, and one chief who had lost a great pile of presents, and who got nothing at all by the second division, stalked solemnly out of the tent, through not until Three Bulls had tossed the plugs of tobacco to all the men around the circle, precisely as he might have thrown bones to dogs, but always observing a certain order in making each round with the plugs. All were thus served according to their rank. Then Three Bulls rummaged with one hand behind him in the grass, and fetched forward a great pipe with a stone bowl and wooden handle—a sort of chopping-block of wood—and a large long-bladed knife. Taking a plug of tobacco in one hand and the knife in the other, he pared off enough tobacco to fill the pipe. Then he filled it, and passed it, stem foremost, to a young man on the left-hand side of the tepee. The superior chiefs all sat on the right-hand side. The young man knew that he had been chosen to perform the menial act of lighting the pipe, and he lighted it, pulling two or three whiffs of smoke to insure a good coal of fire in it before passing it back —through why it was not considered a more menial task to cut the tobacco and fill the pipe than to light it I don't know.

Three Bulls puffed the pipe for a moment, and then turning the stem from him, pointed it at the chief next in importance, and to that personage the symbol of peace was passed from hand to hand. When that chief had drawn a few whiffs, he sent the pipe back to Three Bulls, who then indicated to whom it should go next. Thus it went dodging about the circle like a marble on a bagatelle board.

When it came to me, I hesitated a moment whether or not to smoke it, but the desire to be polite outweighed any other prompting, and I sucked the pipe until some of the Indians cried out that I was "a good fellow."

While all smoked and many talked, I noticed that Three Bulls sat upon a soft seat formed of his blanket, at one end of which was one of those wickerwork contrivances, like a chair back, upon which Indians lean when seated upon the ground. I noticed also that one harsh criticism passed upon Three Bulls was just; that was that when he spoke, others might interrupt him. It was said that even women "talked back" to him at times when he was haranguing his people. Since no one spoke when Crowfoot talked, the comparison between him and his predecessor was injurious to him; but it was Crowfoot who named Three Bulls for the chieftainship. Besides, Three Bulls had the largest following (under that of the too aged Old Sun), and was the most generous chief and ablest politician of all. Then, again, the Government supported him with whatever its influence amounted to. This was because Three Bulls favored agricultural employment for the tribe, and was himself cultivating a patch of potatoes. He was in many other ways the man to lead in the new era, as Crowfoot had been for the era that was past.

When we retired from the presence of the chief, I asked Mr. L'Hereux how he had dared to take back the presents made to the Indians and then distribute them differently. The queer Frenchman said, in his indescribably confident, jaunty way:

"Why, dat is how you mus' do wid dose Engine. Nevaire ask one of dose Engine anyt'ing, but do dose t'ing which are right, and at de same time mak explanashion what you are doing. Den dose Engine can say no t'ing 'tall. But if you first make explanashion and den try to do somet'ng, you will find one grand trouble. Can you explain dis and dat to one hive of de bees? Well, de hive of de bee is like dose Engine if you shall talk widout de promp' action."

He said, later on, "Dose Engine are children, and mus' not haf consideration like mans and women."

The news of our generosity ran from tent to tent, and the Black Soldier band sent out a herald to cry the news that a war-dance was to be held immediately. As immediately means to the Indian mind an indefinite and very enduring period, I amused myself by poking about the village, in tents and among groups of men or women, wherever chance led me. The herald rode from side to side of the enclosure, yelling like a New York fruit peddler. He was mounted on a bay pony, and was fantastically costumed with feathers and war-paint. Of course every man, woman, and child who had been in-doors, so to speak, now came out of the tepees, and a mighty bustle enlivened the scene. The worst thing about the camp was the abundance of snarling cur-dogs. It was not safe to walk about the camp without a cane or whip, on account of these dogs.

The Blackfeet are poor enough, in all conscience, from nearly every stand-point from which we judge civilized communities, but their tribal possessions in-

INDIAN MOTHER AND BOY

clude several horses to each head of a family; and though the majority of their ponies would fetch no more than $20 apiece out there, even this gives them more wealth per capita than many civilized peoples can boast. They have managed, also, to keep much of the savage paraphernalia of other days in the form of buckskin clothes, elaborate bead-work, eagle head-dresses, good guns, and the outlandish adornments of their chiefs and medicine-men. Hundreds of miles from any except such small and distant towns as Calgary and Medicine Hat, and kept on the reserve as much as possible, there has come to them less damage by whiskey and white men's vices than perhaps most other tribes have suffered. Therefore it was still possible for me to see in some tents the squaws at work painting the clan signs on stretched skins, and making bead-work for moccasins, pouches, "chaps," and the rest. And in one tepee I found a young and rather pretty girl wearing a suit of buckskin, such as Cooper and all the past historians of the Indian knew as the conventional every-day attire of the red-skin. I say I saw the girl in a tent, but, as a matter of fact, she passed me out-of-doors, and with true feminine art managed to allow her blanket to fall open for just the instant it took to disclose the precious dress beneath it. I asked to be taken into the tent to which she went, and there, at the interpreter's request, she threw off her blanket, and stood, with a little display of honest coyness, dressed like the traditional and the theatrical belle of the wilderness. The soft yellowish leather, the heavy fringe upon the arms, seams, and edges of the garment, her beautiful beaded leggings and moccasins, formed so many parts of a very charming picture. For herself, her face was comely, but her figure was—an Indian's. The figure of the typical Indian woman shows few graceful curves.

The reader will inquire whether there was any real beauty, as we judge it, among these Indians. Yes, there was; at least there were good looks if there was not beauty. I saw perhaps a dozen fine-looking men, half a dozen attractive girls, and something like a hundred children of varying degrees of comeliness — pleasing, pretty, or beautiful. I had some jolly romps with the children, and so came to know that their faces and arms met my touch with the smoothness and softness of the flesh of our own little ones at home. I was surprised at this; indeed, the skin of the boys was of the texture of velvet. The madcap urchins, what riotous fun they were having! They flung arrows and darts, ran races and wrestled, and in some of their play they fairly swarmed all over one another, until at times one lad would be buried in the thick of a writhing mass of legs and arms several feet in depth. Some of the boys wore only "G-strings" (as, for some reason, the breech-clout is commonly called on the prairie), but others were wrapped in old blankets, and the larger ones were already wearing the Blackfoot plume-lock, or tuft of hair tied and trained to stand erect above the forehead. The babies within the tepees were clad only in their complexions.

The result of an hour of waiting on our part and of yelling on the part of the herald resulted in a war-dance, not very different in itself from the dances we have most of us seen at Wild West shows. An immense tomtom as big as the largest-sized bass-drum was set up between four poles, around which colored cloths were wrapped, and from the tops of which the same gay stuff floated on the wind in bunches of party-colored ribbons. Around this squatted four young braves, who pounded the drum-head and chanted a tune, which rose and fell between the shrillest and the deepest notes, but which consisted of simple monosyllabic sounds repeated thousands of times. The interpreter said that originally the Indians had words to their songs, but these were forgotten no man knows when, and only the so-called tunes (and the tradition that there once were words for them) are perpetuated. At all events, the four braves beat the drum and chanted, until presently a young warrior, hideous with war-paint, and carrying a shield and a tomahawk, came out of a tepee and be-

OPENING OF THE SOLDIER CLAN DANCE

gan the dancing. It was the stiff-legged hopping, first on one foot and then on the other, which all savages appear to deem the highest form the terpsichorean art can take. In the course of a few circles around the tomtom he began shouting of valorous deeds he never had performed, for he was too young to have ridden after buffalo or into battle. Presently he pretended to see upon the ground something at once fascinating and awesome. It was the trail of the enemy. Then he danced furiously and more limberly, tossing his head back, shaking his hatchet and many-tailed shield high aloft, and yelling that he was following the foe, and would not rest while a skull and a scalp-lock remained in conjunction among them. He was joined by three others, and all danced and yelled like madmen. At the last the leader came to a sort of standard made of a stick and some cloth, tore it out from where it had been thrust in the ground, and holding it far above his head, pranced once around the circle, and thus ended the dance.

The novelty and interest in the celebration rested in the surroundings—the great circle of tepees; the braves in their blankets stalking hither and thither; the dogs, the horses, the intrepid riders, dashing across the view. More strange still was the solemn line of the medicine-men, who, for some reason not explained to me, sat in a row with their backs to the dancers a city block away, and crooned a low guttural accompaniment to the tomtom. But still more interesting were the boys, of all grades of childhood, who looked on, while not a woman remained in sight. The larger boys stood about in groups, watching the spectacle with eyes afire with admiration, but the little fellows had flung themselves on their stomachs in a row, and were supporting their chubby faces upon their little brown hands, while their elbows rested on the grass, forming a sort of orchestra row of Lilliputian spectators.

We arranged for a great spectacle to be gotten up on the next afternoon, and were promised that it should be as notable for the numbers participating in it and for the trappings to be displayed as any the Blackfeet had ever given upon their reserve. The Indians spent the entire night in carousing over the gift of tea, and we knew that if they were true to most precedents they would brew and drink every drop of it. Possibly some took it with an admixture of tobacco and wild currant to make them drunk, or, in reality, very sick—which is much the same thing to a reservation Indian. The compounds which the average Indian will swallow in the hope of imitating the effects of whiskey are such as to tax the credulity of those who hear of them. A certain patent "painkiller" ranks almost as high as whiskey in their estimation; but Worcestershire sauce and gunpowder, or tea, tobacco, and wild currant, are not at all to be despised when alcohol, or the money to get it with, is wanting. I heard a characteristic story about these red men while I was visiting them. All who are familiar with them know that if medicine is given them to take in small portions at certain intervals they are morally sure to swallow it all at once, and that the sicker it makes them, the more they will value it. On the Blackfoot Reserve, only a short time

SKETCH IN THE SOLDIER CLAN DANCE

ago, our gentle and insinuating Sedlitz-powders were classed as children's stuff, but now they have leaped to the front rank as powerful medicines. This is because some white man showed the Indian how to take the soda and magnesia first, and then swallow the tartaric acid. They do this, and when the explosion follows, and the gases burst from their mouths and noses, they pull themselves together and remark, "Ugh! him heap good."

On the morning of the day of the great spectacle I rode with Mr. Begg over to the ration-house to see the meat distributed. The dust rose in clouds above all the trails as the cavalcade of men, women, children, travoises and dogs, approached the station. Men were few in the disjointed lines; most of them sent

their women or children. All rode astraddle, some
on saddles and some bareback. As all urged their
horses in the Indian fashion, which is to whip them
unceasingly, and prod them constantly with spurless
heels, the bobbing movement of the riders' heads and
the gymnastics of their legs produced a queer scene.
Here and there a travois was trailed along by a horse
or a dog, but the majority of the pensioners were
content to carry their meat in bags or otherwise upon
their horses. While the slaughtering went on, and
after that, when the beef was being chopped up into
junks, I sat in the meat-contractor's office, and saw
the bucks, squaws, and children come, one after
another, to beg. I could not help noticing that all
were treated with marked and uniform kindness, and
I learned that no one ever struck one of the Indians,
or suffered himself to lose his temper with them. A
few of the men asked for blankets, but the squaws
and the children wanted soap. It was said that when
they first made their acquaintance with this symbol
of civilization they mistook it for an article of diet,
but that now they use it properly and prize it. When
it was announced that the meat was ready, the butch-
ers threw open an aperture in the wall of the ration-
house, and the Indians huddled before it as if they
had flung themselves against the house in a mass.
I have seen boys do the same thing at the opening
of a ticket window for the sale of gallery seats in a
theatre. There was no fighting or quarrelling, but
every Indian pushed steadily and silently with all his
or her might. When one got his share he tore himself away from the crowd as briers are pulled out of

hairy cloth. They are a hungry and an economical people. They bring pails for the beef blood, and they carry home the hoofs for jelly. After a steer has been butchered and distributed, only his horns and his paunch remain.

The sun blazed down on the great camp that afternoon and glorified the place so that it looked like a miniature Switzerland of snowy peaks. But it was hot, and blankets were stretched from the tent tops, and the women sat under them to catch the air and escape the heat. The salaried native policeman of the reserve, wearing a white stove-pipe hat with feathers, and a ridiculous blue coat, and Heaven alone knows what other absurdities, rode around, boasting of deeds he never performed, while a white cur made him all the more ridiculous by chasing him and yelping at his horse's tail.

And then came the grand spectacle. The vast plain was forgotten, and the great campus within the circle of tents was transformed into a theatre. The scene was a setting of white and red tents that threw their clear-cut outlines against a matchless blue sky. The audience was composed of four white men and the Indian boys, who were flung about by the startled horses they were holding for us. The players were the gorgeous cavalrymen of nature, circling before their women and old men and children, themselves plumed like unheard-of tropical birds, the others displaying the minor splendor of the kaleidoscope. The play was "The Pony War-dance, or the Departure for Battle." The acting was fierce; not like the conduct of a mimic battle on our stage, but per-

formed with the desperate zest of men who hope for
distinction in war, and may not trifle about it. It
had the earnestness of a challenged man who tries
the foils with a tutor. It was impressive, inspiring,
at times wildly exciting.

There were threescore young men in the brilliant
cavalcade. They rode horses that were as wild as
themselves. Their evolutions were rude, but magnificent. Now they dashed past us in single file, and
next they came helter-skelter, like cattle stampeding.
For a while they rode around and around, as on a
race-course, but at times they deserted the enclosure,
parted into small bands, and were hidden behind the
curtains of their own dust, presently to reappear with
a mad rush, yelling like maniacs, firing their pieces,
and brandishing their arms and their finery wildly on
high. The orchestra was composed of seven tom-toms that had been dried taut before a camp fire.
The old men and the chiefs sat in a semicircle behind the drummers on the ground.

All the tribal heirlooms were in the display, the
cherished gewgaws, trinkets, arms, apparel, and finery
they had saved from the fate of which they will not
admit they are themselves the victims. I never saw
an old-time picture of a type of savage red man or of
an extravagance of their costuming that was not revived in this spectacle. It was as if the plates in my
old school-books and novels and tales of adventure
were all animated and passing before me. The traditional Indian with the eagle plumes from crown to
heels was there; so was he with the buffalo horns
growing out of his skull; so were the idyllic braves

in yellow buckskin fringed at every point. The shining bodies of men, bare naked, and frescoed like a Bowery bar-room, were not lacking; neither were those who wore masses of splendid embroidery with colored beads. But there were as many peculiar costumes which I never had seen pictured. And not any two men or any two horses were alike. As barber poles are covered with paint, so were many of these choice steeds of the nation. Some were spotted all over with daubs of white, and some with every color obtainable. Some were branded fifty times with the white hand, the symbol of peace, but others bore the red hand and the white hand in alternate prints. There were horses painted with the figures of horses and of serpents and of foxes. To some saddles were affixed colored blankets or cloths that fell upon the ground or lashed the air, according as the horse cantered or raced. One horse was hung all round with great soft woolly tails of some white material. Sleigh-bells were upon several.

Only half a dozen men wore hats—mainly cowboy hats decked with feathers. Many carried rifles, which they used with one hand. Others brought out bows and arrows, lances decked with feathers or ribbons, poles hung with colored cloths, great shields brilliantly painted and fringed. Every visible inch of each warrior was painted, the naked ones being ringed, streaked, and striped from head to foot. I would have to catalogue the possessions of the whole nation to tell all that they wore between the brass rings in their hair and the cartridge-belts at their waists, and thus down to their beautiful moccasins.

Two strange features further distinguished their pageant. One was the appearance of two negro minstrels upon one horse. Both had blackened their faces and hands; both wore old stove-pipe hats and queer long-tailed white men's coats. One wore a huge false white mustache, and the other carried a coal-scuttle. The women and children roared with laughter at the sight. The two comedians got down from their horse, and began to make grimaces, and to pose this way and that, very comically. Such a performance had never been seen on the reserve before. No one there could explain where the men had seen negro minstrels. The other unexpected feature required time for development. At first we noticed that two little Indian boys kept getting in the way of the riders. As we were not able to find any fixed place of safety from the excited horsemen, we marvelled that these children were permitted to risk their necks.

Suddenly a hideously-painted naked man on horseback chased the little boys, leaving the cavalcade, and circling around the children. He rode back into the ranks, and still they loitered in the way. Then around swept the horsemen once more, and this time the naked rider flung himself from his horse, and seizing one boy and then the other, bore each to the ground, and made as if he would brain them with his hatchet and lift their scalps with his knife. The sight was one to paralyze an on-looker. But it was only a theatrical performance arranged for the occasion. The man was acting over again the proudest of his achievements. The boys played the

THROWING THE SNOW SNAKE

parts of two white men whose scalps now grace his tepee and gladden his memory.

For ninety minutes we watched the glorious riding, the splendid horses, the brilliant trappings, and the paroxysmal fervor of the excited Indians. The earth trembled beneath the dashing of the riders; the air palpitated with the noise of their war-cries and bells. We could have stood the day out, but we knew the players were tired, and yet would not cease till we withdrew. Therefore we came away.

We had enjoyed a never-to-be-forgotten privilege. It was as if we had seen the ghosts of a dead people ride back to parody scenes in an era that had vanished. It was like the rising of the curtain, in response to an "encore," upon a drama that has been played. It was as if the sudden up-flashing of a smouldering fire lighted, once again and for an instant, the scene it had ceased to illumine.

III

A FAMOUS MISSIONARY

THE former chief of the Blackfeet—Crowfoot—and Father Lacombe, the Roman Catholic missionary to the tribe, were the most interesting and among the most influential public characters in the newer part of Canada. They had much to do with controlling the peace of a territory the size of a great empire.

The chief was more than eighty years old; the priest is a dozen years younger; and yet they represented in their experiences the two great epochs of life on this continent—the barbaric and the progressive. In the chief's boyhood the red man held undisputed sway from the Lakes to the Rockies. In the priest's youth he led, like a scout, beyond the advancing hosts from Europe. But Father Lacombe came bearing the olive branch of religion, and he and the barbarian became fast friends, intimates in a companionship as picturesque and out of the common as any the world could produce.

There is something very strange about the relations of the French and the French half-breeds with the wild men of the plains. It is not altogether necessary that the Frenchman should be a priest, for I have heard of French half-breeds in our Territories who showed again and again that they could make their way through bands of hostiles in perfect safety, though knowing nothing of the language of the tribes there in war-paint. It is most likely that their swarthy skins and black hair, and their knowledge of savage ways aided them. But when not even a French half-breed has dared to risk his life among angry Indians, the French missionaries went about their duty fearlessly and unscathed. There was one, just after the dreadful massacre of the Little Big Horn, who built a cross of rough wood, painted it white, fastened it to his buck-board, and drove through a country in which a white man with a pale face and blond hair would not have lived two hours.

It must be remembered that in a vast region of country the French priest and *voyageur* and *coureur des bois* were the first white men the Indians saw, and while the explorers and traders seldom quarrelled with the red men or offered violence to them, the priests never did. They went about like women or children, or, rather, like nothing else than priests. They quickly learned the tongues of the savages, treated them fairly, showed the sublimest courage, and acted as counsellors, physicians, and friends. There is at least one brave Indian fighter in our army who will state it as his belief that if all the white men had done thus we would have had but little trouble with our Indians.

Father Lacombe was one of the priests who threaded the trails of the North-western timber land and the Far Western prairie when white men were very few indeed in that country, and the only settlements were those that had grown around the frontier forts and the still earlier mission chapels. For instance, in 1849, at twenty-two years of age, he slept a night or two where St. Paul now weights the earth. It was then a village of twenty-five log-huts, and where the great building of the St. Paul *Pioneer Press* now stands, then stood the village chapel. For two years he worked at his calling on either side of the American frontier, and then was sent to what is now Edmonton, in that magical region of long summers and great agricultural capacity known as the Peace River District, hundreds of miles north of Dakota and Idaho. There the Rockies are broken and lowered, and the warm Pacific winds have rendered the region warmer than the land far to the south of it. But Father Lacombe went farther—400 miles north to Lake Labiche. There he found what he calls a fine colony of half-breeds. These were dependants of the Hudson Bay Company—white men from England, France, and the Orkney Islands, and Indians and half-breeds and their children. The visits of priests were so infrequent that in the intervals between them the white men and Indian women married one another, not without formality and the sanction of the colony, but without waiting for the ceremony of the Church. Father Lacombe was called upon to bless and solemnize many such matches, to baptize many children, and to teach and preach what scores knew but vaguely or not at all.

In time he was sent to Calgary, in the province of Alberta. It is one of the most bustling towns in the Dominion, and the biggest place west of Winnipeg. Alberta is north of our Montana, and is all prairie-land; but from Father Lacombe's parsonage one sees the snow-capped Rockies, sixty miles away, lying above the horizon like a line of clouds tinged with the delicate hues of mother-of-pearl in the sunshine. Calgary was a mere post in the wilderness for years after the priest went there. The buffaloes roamed the prairie in fabulous numbers, the Indians used the bow and arrow in the chase, and the maps we studied at the time showed the whole region enclosed in a loop, and marked "Blackfoot Indians." But the other Indians were loath to accept this disposition of the territory as final, and the country thereabouts was an almost constant battle-ground between the Blackfoot nation of allied tribes and the Sioux, Crows, Flatheads, Crees, and others.

The good priest—for if ever there was a good man Father Lacombe is one—saw fighting enough, as he roamed with one tribe and the other, or journeyed from tribe to tribe. His mission led him to ignore tribal differences, and to preach to all the Indians of the plains. He knew the chiefs and headmen among them all, and so justly did he deal with them that he was not only able to minister to all without attracting the enmity of any, but he came to wield, as he does to-day, a formidable power over all of them.

He knew old Crowfoot in his prime, and as I saw them together they were like bosom friends. Together they had shared dreadful privation and survived frightful winters and storms. They had gone side by side through savage battles, and each respected and loved the other. I think I make no mistake in saying that all through his reign Crowfoot was the greatest Indian monarch in Canada; possibly no tribe in this country was stronger in numbers during the last decade or two. I have never seen a nobler-looking Indian or a more king-like man. He was tall and straight, as slim as a girl, and he had the face of an eagle or of an ancient Roman. He never troubled himself to learn the English language; he had little use for his own. His grunt or his "No" ran all through his tribe. He never shared his honors with a squaw. He died an old bachelor, saying, wittily, that no woman would take him.

It must be remembered that the degradation of the Canadian Indian began a dozen or fifteen years later than that of our own red men. In both countries the railroads were indirectly the destructive agents, and Canada's great transcontinental line is a new institution. Until it belted the prairie the other day the Blackfoot Indians led very much the life of their fathers, hunting and trading for the whites, to be sure, but living like Indians, fighting like Indians, and dying like them. Now they don't fight, and they live and die like dogs. Amid the old conditions lived Crowfoot—a haughty, picturesque, grand old savage. He never rode or walked without his headmen in his retinue, and when he wished to exert his authority, his apparel was royal indeed. His coat of gaudy bead-work was a splendid garment, and weighed a dozen pounds. His leg-gear was just as fine; his moccasins would fetch fifty dollars in any city to-day. Doubtless he thought his hat was quite as impressive and king-like, but to a mere scion of effeminate civilization it looked remarkably like an

extra tall plug hat, with no crown in the top and a
lot of crows' plumes in the band. You may be sure
his successor wears that same hat to-day, for the Indians revere the "state hat" of a brave chief, and look
at it through superstitious eyes, so that those queer
hats (older tiles than ever see the light of St. Patrick's
Day) descend from chief to chief, and are hallowed.

But Crowfoot died none too soon. The history of
the conquest of the wilderness contains no more pathetic story than that of how the kind old priest,
Father Lacombe, warned the chief and his lieutenants against the coming of the pale-faces. He went
to the reservation and assembled the leaders before
him in council. He told them that the white men
were building a great railroad, and in a month their
workmen would be in that virgin country. He told
the wondering red men that among these laborers
would be found many bad men seeking to sell whiskey, offering money for the ruin of the squaws.
Reaching the greatest eloquence possible for him,
because he loved the Indians and doubted their
strength, he assured them that contact with these
white men would result in death, in the destruction
of the Indians, and by the most horrible processes of
disease and misery. He thundered and he pleaded.
The Indians smoked and reflected. Then they
spoke through old Crowfoot:

"We have listened. We will keep upon our reservation. We will not go to see the railroad."

But Father Lacombe doubted still, and yet more
profoundly was he convinced of the ruin of the tribe
should the "children," as he sagely calls all Indians,

disobey him. So once again he went to the reserve, and gathered the chief and the headmen, and warned them of the soulless, diabolical, selfish instincts of the white men. Again the grave warriors promised to obey him.

The railroad laborers came with camps and money and liquors and numbers, and the prairie thundered the echoes of their sledge-hammer strokes. And one morning the old priest looked out of the window of his bare bedroom and saw curling wisps of gray smoke ascending from a score of tepees on the hill beside Calgary.* Angry, amazed, he went to his doorway and opened it, and there upon the ground sat some of the headmen and the old men, with bowed heads, ashamed. Fancy the priest's wrath and his questions! Note how wisely he chose the name of children for them, when I tell you that their spokesman at last answered with the excuse that the buffaloes were gone, and food was hard to get, and the white men brought money which the squaws could get. And what is the end? There are always tepees on the hills now beside every settlement near the Blackfoot reservation. And one old missionary lifted his trembling forefinger towards the sky, when I was there, and said: "Mark me. In fifteen years there will not be a full-blooded Indian alive on the Canadian prairie—not one."

Through all that revolutionary railroad building and the rush of new settlers, Father Lacombe and

* Since this was written Father Lacombe's work has been continued at Fort McLeod in the same province as Calgary. In this smaller place he finds more time for his literary pursuits.

Crowfoot kept the Indians from war, and even from depredations and from murder. When the half-breeds arose under Riel, and every Indian looked to his rifle and his knife, and when the mutterings that preface the war-cry sounded in every lodge, Father Lacombe made Crowfoot pledge his word that the Indians should not rise. The priest represented the Government on these occasions. The Canadian statesmen recognize the value of his services. He is the great authority on Indian matters beyond our border; the ambassador to and spokesman for the Indians.

But Father Lacombe is more than that. He is the deepest student of the Indian languages that Canada possesses. The revised edition of Bishop Barager's *Grammar of the Ochipwe Language* bears these words upon its title-page: "Revised by the Rev. Father Lacombe, Oblate Mary Immaculate, 1878." He is the author of the authoritative *Dictionnaire et Grammaire de la Langue Crise*, the dictionary of the Cree dialect published in 1874. He has compiled just such another monument to the Blackfoot language, and will soon publish it, if he has not done so already. He is in constant correspondence with our Smithsonian Institution; he is famous to all who study the Indian; he is beloved or admired throughout Canada.

His work in these lines is labor of love. He is a student by nature. He began the study of the Algonquin language as a youth in older Canada, and the tongues of many of these tribes from Labrador to Athabasca are but dialects of the language of the

FATHER LACOMBE HEADING THE INDIANS

great Algonquin nation—the Algic family. He told me that the white man's handling of Indian words in the nomenclature of our cities, provinces, and States is as brutal as anything charged against the savages. Saskatchewan, for instance, means nothing. "Kissiskatchewan" is the word that was intended. It means "rapid current." Manitoba is senseless, but "Manitowapa" (the mysterious strait) would have been full of local import. However, there is no need to sadden ourselves with this expert knowledge. Rather let us be grateful for every Indian name with which we have stamped individuality upon the map of the world, be it rightly or wrongly set forth.

It is strange to think of a scholar and a priest amid the scenes that Father Lacombe has witnessed. It was one of the most fortunate happenings of my life that I chanced to be in Calgary and in the little mission beside the chapel when Chief Crowfoot came to pay his respects to his old black-habited friend. Anxious to pay the chief such a compliment as should present the old warrior to me in the light in which he would be most proud to be viewed, Father Lacombe remarked that he had known Crowfoot when he was a young man and a mighty warrior. The old copper-plated Roman smiled and swelled his chest when this was translated. He was so pleased that the priest was led to ask him if he remembered one night when a certain trouble about some horses, or a chance duel between the Blackfoot tribe and a band of its enemies, led to a midnight attack. If my memory serves me, it was the Bloods (an allied part of the Blackfoot nation) who picked this quarrel. The chief grinned and grunted wonderfully as the priest spoke. The priest asked if he remembered how the Bloods were routed. The chief grunted even more emphatically. Then the priest asked if the chief recalled what a pickle he, the priest, was in when he found himself in the thick of the fight. At that old Crowfoot actually laughed.

After that Father Lacombe, in a few bold sentences, drew a picture of the quiet, sleep-enfolded camp of the Blackfoot band, of the silence and the darkness. Then he told of a sudden musket-shot; then of the screaming of the squaws, and the barking of the dogs, and the yelling of the children, of the general hubbub and confusion of the startled camp. The cry was everywhere "The Bloods! the Bloods!" The enemy shot a fusillade at close quarters into the Blackfoot camp, and the priest ran out towards the blazing muskets, crying that they must stop, for he, their priest, was in the camp. He shouted his own name, for he stood towards the Bloods precisely as he did towards the Blackfoot nation. But whether the Bloods heard him or not, they did not heed him. The blaze of their guns grew stronger and crept nearer. The bullets whistled by. It grew exceedingly unpleasant to be there. It was dangerous as well. Father Lacombe said that he did all he could to stop the fight, but when it was evident that his behavior would simply result in the massacre of his hosts and of himself in the bargain, he altered his cries into military commands. "Give it to 'em!" he screamed. He urged Crowfoot's braves to return two shots for every one from the enemy.

He took command, and inspired the bucks with double valor. They drove the Bloods out of reach and hearing.

All this was translated to Crowfoot—or Saponaxitaw, for that was his Indian name—and he chuckled and grinned, and poked the priest in the side with his knuckles. And good Father Lacombe felt the magnetism of his own words and memory, and clapped the chief on the shoulder, while both laughed heartily at the climax, with the accompanying mental picture of the discomfited Bloods running away, and the clergyman ordering their instant destruction.

There may not be such another meeting and rehearsal on this continent again. Those two men represented the passing and the dominant races of America; and yet, in my view, the learned and brave and kindly missionary is as much a part of the dead past as is the royalty that Crowfoot was the last to represent.

IV

ANTOINE'S MOOSE-YARD

IT was the night of a great dinner at the club. Whenever the door of the banqueting hall was opened, a burst of laughter or of applause disturbed the quiet talk of a few men who had gathered in the reading-room—men of the sort that extract the best enjoyment from a club by escaping its functions, or attending them only to draw to one side its choicest spirits for never-to-be-forgotten talks before an open fire, and over wine and cigars used sparingly.

"I'm tired," an artist was saying—"so tired that I have a horror of my studio. My wife understands my condition, and bids me go away and rest."

"That is astonishing," said I; "for, as a rule, neither women nor men can comprehend the fatigue that seizes an artist or writer. At most of our homes there comes to be a reluctant recognition of the fact that we say we are tired, and that we persist in the assumption by knocking off work. But human fatigue is measured by the mile of walking, or the cords of firewood that have been cut, and the world will always hold that if we have not hewn wood or tramped all day, it is absurd for us to talk of feeling tired. We cannot alter this; we are too few."

"Yes," said another of the little party. "The world shares the feeling of the Irishman who saw a very large, stout man at work at reporting in a courtroom. 'Faith!' said he, 'will ye look at the size of that man—to be airning his living wid a little pincil?' The world would acknowledge our right to feel tired if we used crow-bars to write or draw with; but pencils! pshaw! a hundred weigh less than a pound."

"Well," said I, "all the same, I am so tired that my head feels like cork; so tired that for two days I have not been able to summon an idea or turn a sentence neatly. I have been sitting at my desk writing wretched stuff and tearing it up, or staring blankly out of the window."

"Glorious!" said the artist, startling us all with his vehemence and inapt exclamation. "Why, it is providential that I came here to-night. If that's the way you feel, we are a pair, and you will go with me and rest. Do you hunt? Are you fond of it?"

"I know all about it," said I, "but I have not definitely determined whether I am fond of it or not. I have been hunting only once. It was years ago, when I was a mere boy. I went after deer with a poet, an editor, and a railroad conductor. We journeyed to a lovely valley in Mifflin County, Pennsylvania, and put ourselves in the hands of a man seven feet high, who had a flintlock musket a foot taller than himself, and a wife who gave us saleratus bread and a bowl of pork fat for supper and breakfast. We were not there at dinner. The man stationed us a mile apart on what he said were the paths, or "runways," the deer would take. Then he went to stir the game up with his dogs. There he left us from sunrise till supper, or would have left us had we not with great difficulty found one another, and enjoyed the exquisite woodland quiet and light and shade together, mainly flat on our backs, with the white sails of the sky floating in an azure sea above the reaching fingers of the tree-tops. The editor marred the occasion with an unworthy suspicion that our hunter was at the village tavern picturing to his cronies what simple donkeys we were, standing a mile apart in the forsaken woods. But the poet said something so pregnant with philosophy that it always comes back to me with the mention of hunting. 'Where is your gun?' he was asked, when we came upon him, pacing the forest path, hands in pockets, and no weapon in sight. 'Oh, my gun?' he repeated. 'I don't know. Somewhere in among those trees. I covered it with leaves so as not to see it. After this, if I go hunting again, I shall not take a gun. It is very cold and heavy, and more or less dangerous in the bargain. You never use it, you know. I go hunting every few years, but I never yet have had to fire my gun, and I begin to see that it is only brought along in deference to a tradition descending from an era when men got something more than fresh air and scenery on a hunting trip.'"

The others laughed at my story, but the artist regarded me with an expression of pity. He is a famous hunter—a genuine, devoted hunter—and one

THE HOTEL—LAST SIGN OF CIVILIZATION

might almost as safely speak a light word of his relations as of his favorite mode of recreation.

"Fresh air!" said he; "scenery! Humph! Your poet would not know which end of a gun to aim with. I see that you know nothing at all about hunting, but I will pay you the high compliment of saying that I can make a hunter of you. I have always insisted heretofore that a hunter must begin in boyhood; but never mind, I'll make a hunter of you at thirty-six. We will start to-morrow morning for Montreal, and in twenty-four hours you shall be in the greatest sporting region in America, incomparably the greatest hunting district. It is great because Americans do not know of it, and because it has all of British America to keep it supplied with game. Think of it! In twenty-four hours we shall be tracking moose near Hudson Bay, for Hudson Bay is not much farther from New York than Chicago—another fact that few persons are aware of."

Environment is a positive force. We could feel

that we were disturbing what the artist would call
"the local tone," by rushing through the city's streets
next morning with our guns slung upon our backs.
It was just at the hour when the factory hands and
the shop-girls were out in force, and the juxtaposition of those elements of society with two portly
men bearing guns created a positive sensation. In
the cars the artist held forth upon the terrors of the
life upon which I was about to venture. He left
upon my mind a blurred impression of sleeping out-of-doors, like human cocoons, done up in blankets,
while the savage mercury lurked in unknown depths
below the zero mark. He said the camp-fire would
have to be fed every two hours of each night, and he
added, without contradiction from me, that he supposed he would have to perform this duty, as he was
accustomed to it. Lest his forecast should raise my
anticipation of pleasure extravagantly, he added that
those hunters were fortunate who had fires to feed;
for his part he had once walked around a tree stump
a whole night to keep from freezing. He supposed
that we would perform our main journeying on snow-shoes, but how we should enjoy that he could not
say, as his knowledge of snow-shoeing was limited.

At this point the inevitable offspring of fate, who
is always at a traveller's elbow with a fund of alarming information, cleared his throat as he sat opposite
us, and inquired whether he had overheard that we
did not know much about snow-shoes. An interesting fact concerning them, he said, was that they
seemed easy to walk with at first, but if the learner
fell down with them on it usually needed a consider-

able portion of a tribe of Indians to put him back on his feet. Beginners only fell down, however, in attempting to cross a log or stump, but the forest where we were going was literally floored with such obstructions. The first day's effort to navigate with snow-shoes, he remarked, is usually accompanied by a terrible malady called *mal de raquette*, in which the cords of one's legs become knotted in great and excruciatingly painful bunches. The cure for this is to "walk it off the next day, when the agony is yet more intense than at first." As the stranger had reached his destination, he had little more than time to remark that the moose is an exceedingly vicious animal, invariably attacking all hunters who fail to kill him with the first shot. As the stranger stepped upon the car platform he let fall a simple but touching eulogy upon a dear friend who had recently lost his life by being literally cut in two, lengthwise, by a moose that struck him on the chest with its rigidly stiffened fore-legs. The artist protested that the stranger was a sensationalist, unsupported by either the camp-fire gossip or the literature of hunters. Yet one man that night found his slumber tangled with what the garrulous alarmist had been saying.

In Montreal one may buy clothing not to be had in the United States: woollens thick as boards, hosiery that wards off the cold as armor resists missiles, gloves as heavy as shoes, yet soft as kid, fur caps and coats at prices and in a variety that interest poor and rich alike, blanket suits that are more picturesque than any other masculine garment worn north of the city of Mexico, tuques, and moccasins, and, indeed, so many sorts of clothing we Yankees know very little of (though many of us need them) that at a glance we say the Montrealers are foreigners. Montreal is the gayest city on this continent, and I have often thought that the clothing there is largely responsible for that condition.

A New Yorker disembarking in Montreal in midwinter finds the place inhospitably cold, and wonders how, as well as why, any one lives there. I well remember standing years ago beside a toboggan-slide, with my teeth chattering and my very marrow slowly congealing, when my attention was called to the fact that a dozen ruddy-cheeked, bright-eyed, laughing girls were grouped in snow that reached their knees. I asked a Canadian lady how that could be possible, and she answered with a list of the principal garments those girls were wearing. They had two pairs of stockings under their shoes, and a pair of stockings over their shoes, with moccasins over them. They had so many woollen skirts that an American girl would not believe me if I gave the number. They wore heavy dresses and buckskin jackets, and blanket suits over all this. They had mittens over their gloves, and fur caps over their knitted hoods. It no longer seemed wonderful that they should not heed the cold; indeed, it occurred to me that their bravery amid the terrors of tobogganing was no bravery at all, since a girl buried deep in the heart of such a mass of woollens could scarcely expect damage if she fell from a steeple. When next I appeared out-of-doors I too was swathed in flannel, like a jewel in a box of plush, and from that time out Montreal seemed,

"GIVE ME A LIGHT"

what it really is, the merriest of American capitals. And there I had come again, and was filling my trunk with this wonderful armor of civilization, while the artist sought advice as to which point to enter the wilderness in order to secure the biggest game most quickly.

Mr. W. C. Van Horne, the President of the Canadian Pacific Railroad, proved a friend in need. He dictated a few telegrams that agitated the people of a vast section of country between Ottawa and the Great Lakes. And in the afternoon the answers came flying back. These were from various points where Hudson Bay posts are situated. At one or two the Indian trappers and hunters were all away on their winter expeditions; from another a famous white hunter had just departed with a party of gentlemen. At Mattawa, in Ontario, moose were close at hand and plentiful, and two skilled Indian hunters were just in from a trapping expedition; but the post factor, Mr. Rankin, was sick in bed, and the Indians were on a spree. To Mattawa we decided to go. It is a twelve-hour journey from New York to Montreal, and an eleven-hour journey from Montreal to the heart of this hunters' paradise; so that, had we known at just what point to enter the forest, we could have taken the trail in twenty-four hours from the metropolis, as the artist had predicted.

Our first taste of the frontier, at Peter O'Farrall's Ottawa Hotel, in Mattawa, was delicious in the extreme. O'Farrall used to be game-keeper to the Marquis of Waterford, and thus got "a taste of the quality" that prompted him to assume the position he has chosen as the most lordly hotel-keeper in Canada. We do not know what sort of men own our great New York and Chicago and San Francisco hotels, but certainly they cannot lead more leisurely, complacent lives than Mr. O'Farrall. He has a bartender to look after the male visitors and the bar, and a matronly relative to see to the women and the kitchen, so that the landlord arises when he likes to enjoy each succeeding day of ease and prosperity. He has been known to exert himself, as when he chased a man who spoke slightingly of his liquor. And he was momentarily ruffled at the trying conduct of the artist on this hunting trip. The artist could not find his overcoat, and had the temerity to refer the matter to Mr. O'Farrall.

"Sir," said the artist, "what do you suppose has become of my overcoat? I cannot find it anywhere."

"I don't know anything about your botheration overcoat," said Mr. O'Farrall. "Sure, I've throuble enough kaping thrack of me own."

The reader may be sure that O'Farrall's was rightly recommended to us, and that it is a well-managed and popular place, with good beds and excellent fare, and with no extra charge for the delightful addition of the host himself, who is very tall and dignified and humorous, and who is the oddest and yet most picturesque-looking public character in the Dominion. Such an oddity is certain to attract queer characters to his side, and Mr. O'Farrall is no exception to the rule. One of the waiter-girls in the dining-room was found never by any chance to know anything that she was asked about. For instance, she had never heard of Mr. Rankin, the chief man of the place. To every question she made answer, "Sure, there does be a great dale goin' on here and I know nothin' of it." Of her the artist ventured the theory that "she could not know everything on a waiter-girl's salary." John, the bar-tender, was a delightful study. No matter what a visitor laid down in the smoking-room, John picked it up and carried it behind the bar. Every one was continually losing something and searching for it, always to observe that John was able to produce it with a smile and the wise remark that he had taken the lost article and put it away "for fear some one would pick it up." Finally, there was Mr. O'Farrall's dog—a ragged, time-worn, petulant terrier, no bigger than a pint-pot. Mr. O'Farrall nevertheless called him "Fairy," and said he kept him "to protect the village children against wild bears."

I shall never be able to think of Mattawa as it is—a plain little lumbering town on the Ottawa River, with the wreck and ruin of once grand scenery hemming it in on all sides, in the form of ragged mountains literally ravaged by fire and the axe. Hints of it come back to me in dismembered bits that prove it to have been interesting: vignettes of little schoolboys in blanket suits and moccasins, of great-spirited horses forever racing ahead of fur-laden sleighs, and of troops of olive-skinned French-Canadian girls, bundled up from their feet to those mischievous features which shot roguish glances at the artist—the biggest man, the people said, who had ever been seen in Mattawa. But the place will ever yield back to my mind the impression I got of the wonderful preparations that were made for our adventure—preparations that seemed to busy or to interest nearly every one in the village. Our Indians had come in from the Indian village three miles away, and had said they had had enough drink. Mr. John De Sousa, accountant at the post, took charge of them and of us, and the work of loading a great portage sleigh went on apace. The men of sporting tastes came out and lounged in front of the post, and gave helpful advice; the Indians and clerks went to and from the sleigh laden with bags of necessaries; the harness-maker made for us belts such as the lumbermen use to preclude the possibility of incurable strains in the rough life in the wilderness. The help at O'Farrall's assisted in repacking what we needed, so that our trunks and town clothing could be stored. Mr. De Sousa sent messengers hither and thither for essentials not in stock at the post. Some women, even, were set at work to make "neaps" for us, a neap being a sort of slipper or unlaced shoe made of heavy blanketing and worn outside one's stockings, to give added warmth to the feet.

"You see, this is no casual rabbit-hunt," said the artist. The remark will live in Mattawa many a year.

The Hudson Bay Company's posts differ. In the wilderness they are forts surrounded by stockades, but within the boundaries of civilization they are stores. That at Winnipeg is a splendid emporium, while that at Mattawa is like a village store in the United States, except that the top story is laden with guns, traps, snow-shoes, and the skins of wild beasts; while an out-building in the rear is the repository of scores of birch-bark canoes—the carriages of British America. Mr. Rankin, the factor there, lay in a bed of suffering and could not see us. Yet it seemed difficult to believe that we could be made the recipients of greater or more kindly attentions than were lavished upon us by his accountant, Mr. De Sousa.

ANTOINE, FROM LIFE.

He ordered our tobacco ground for us ready for our pipes; selected the finest from among those extraordinary blankets that have been made exclusively for this company for hundreds of years; picked out the largest snow-shoes in his stock; bade us lay aside the gloves we had brought, and take mittens such as he produced, and for which we thanked him in our hearts many times afterwards; planned our outfit of food with the wisdom of an old campaigner; bethought himself to send for baker's bread; ordered high legs sewed on our moccasins—in a word, he made it possible for us to say afterwards that absolutely nothing had been overlooked or slighted in fitting out our expedition.

As I sat in the sleigh, tucked in under heavy skins and leaning at royal ease against other furs that covered a bale of hay, it seemed to me that I had become part of one of such pictures as we all have seen, portraying historic expeditions in Russia or Siberia. We carried fifteen hundred pounds of traps and provisions for camping, stabling, and food for men and beasts. We were five in all—two hunters, two Indians, and a teamster. We set out with the two huge mettlesome horses ahead, the driver on a high seat formed of a second bale of hay, ourselves lolling back under our furs, and the two Indians striding along over the resonant cold snow behind us. It was beginning to be evident that a great deal of effort and machinery was needed to "make a hunter" of a city man, and that it was going to be done thoroughly—two thoughts of a highly flattering nature.

We were now clad for arctic weather, and perhaps nothing except a mummy was ever "so dressed up" as we were. We each wore two pairs of the heaviest woollen stockings I ever saw, and over them ribbed bicycle stockings that came to our knees. Over these in turn were our "neaps," and then our moccasins, laced tightly around our ankles. We had on two suits of flannels of extra thickness, flannel shirts, reefing jackets, and "capeaux," as they call their long-hooded blanket coats, longer than snow-shoe coats. On our heads we had knitted tuques, and on our hands mittens and gloves. We were bound for Antoine's moose-yard, near Crooked Lake.

The explanation of the term "moose-yard" made moose-hunting appear a simple operation (once we were started), for a moose-yard is the feeding-ground of a herd of moose, and our head Indian, Alexandre Antoine, knew where there was one. Each herd or family of these great wild cattle has two such feeding-grounds, and they are said to go alternately from one to the other, never herding in one place two years in succession. In this region of Canada they weigh between 600 and 1200 pounds, and the reader will help his comprehension of those figures by recalling the fact that a 1200-pound horse is a very large one. Whether they desert a yard for twelve months because of the damage they do to the supply of food it offers to them, or whether it is instinctive caution that directs their movements, no one can more than conjecture.

Their yards are always where soft wood is plentiful and water is near, and during a winter they will feed over a region from half a mile to a mile square. The prospect of going directly to the fixed home of a herd of moose almost robbed the trip of that speculative element that gives the greatest zest to hunting. But we knew not what the future held for us. Not even the artist, with all his experience, conjectured what was in store for us. And what was to come began coming almost immediately.

The journey began upon a good highway, over which we slid along as comfortably as any ladies in their carriages, and with the sleigh-bells flinging their cheery music out over a desolate valley, with a leaden river at the bottom, and with small mountains rolling all about. The timber was cut off them, except here and there a few red or white pines that reared their green, brush-like tops against the general blanket of snow. The dull sky hung sullenly above, and now and then a raven flew by, croaking hoarse disapproval of our intrusion. To warn us of what we were to expect, Antoine had made a shy

Indian joke, one of the few I ever heard: "In small little while," said he, "we come to all sorts of a road. Me call it that 'cause you get every sort riding, then you sure be suited."

At five miles out we came to this remarkable highway. It can no more be adequately described here than could the experiences of a man who goes over Niagara Falls in a barrel. The reader must try to imagine the most primitive sort of a highway conceivable—one that has been made by merely felling trees through a forest in a path wide enough for a team and wagon. All the tree stumps were left in their places, and every here and there were rocks; some no larger than a bale of cotton, and some as small as a bushel basket. To add to the other alluring qualities of the road, there were tree trunks now and then directly across it, and, as a further inducement to traffic, the highway was frequently interrupted by "pitch holes." Some of these would be called pitch holes anywhere. They were at points where a rill crossed the road, or the road crossed the corner of a marsh. But there were other pitch holes that any intelligent New Yorker would call ravines or gullies. These were at points where one hill ran down to the water-level and another immediately rose precipitately, there being a watercourse between the two. In all such places there was deep black mud and broken ice. However, these were mere features of the character of this road—a character too profound for me to hope to portray it. When the road was not inclined either straight down or straight up, it coursed along the slanting side of a steep hill, so that a vehicle could keep to it only by falling against the forest at the under side and carroming along from tree to tree.

Such was the road. The manner of travelling it was quite as astounding. For nothing short of what Alphonse, the teamster, did would I destroy a man's character; but Alphonse was the next thing to an idiot. He made that dreadful journey at a gallop! The first time he upset the sleigh and threw me with one leg thigh-deep between a stone and a tree trunk, besides sending the artist flying over my head like a shot from a sling, he reseated himself and remarked: "That makes tree time I upset in dat place. Hi, there! Get up!" It never occurred to him to stop because a giant tree had fallen across the trail. "Look out! Hold tight!" he would call out, and then he would take the obstruc-

THE PORTAGE SLEIGH ON A LUMBER ROAD

tion at a jump. The horses were mammoth beasts, in the best fettle, and the sleigh was of the solidest, strongest pattern. There were places where even Alphonse was anxious to drive with caution. Such were the ravines and unbridged waterways. But one of the horses had cut himself badly in such a place a year before, and both now made it a rule to take all such places flying. Fancy the result! The leap in air, and then the crash of the sled as it landed, the snap of the harness chains, the snorts of the winded beasts, the yells of the driver, the anxiety and nervousness of the passengers!

At one point we had an exciting adventure of a far different sort. There was a moderately good stretch of road ahead, and we invited the Indians to jump in and ride a while. We noticed that they took occasional draughts from a bottle. They finished a full pint, and presently Alexandre produced another and larger phial. Every one knows what a drunken Indian is, and so did we. We ordered the sleigh stopped and all hands out for "a talk." Firmly, but with both power and reason on our side, we demanded a promise that not another drink should be taken, or that the horses be turned towards Mattawa at once. The promise was freely given.

"But what is that stuff? Let me see it," one of the hunters asked.

"It is de 'igh wine," said Alexandre.

"High wine? Alcohol?" exclaimed the hunter, and, impulse being quicker than reason sometimes, flung the bottle high in air into the bush. It was an injudicious action, but both of us at once prepared to defend and re-enforce it, of course. As it happened, the Indians saw that no unkindness or unfairness was intended, and neither sulked nor made trouble afterwards.

We were now deep in the bush. Occasionally we passed "a brulè," or tract denuded of trees, and littered with trunks and tops of trunks rejected by the lumbermen. But every mile took us nearer to the undisturbed primeval forest, where the trees shoot up forty feet before the branches begin. There were no houses, teams, or men. In a week in the bush we saw no other sign of civilization than what we brought or made. All around us rose the motionless regiments of the forest, with the snow beneath them, and their branches and twigs printing lacework on the sky. The signs of game were numerous, and varied to an extent that I never heard of before. There were few spaces of the length of twenty-five feet in which the track of some wild beast or bird did not cross the road. The Indians read this writing in the snow, so that the forest was to them as a book would be to us. "What is that?" "And that?" "And that?" I kept inquiring. The answers told more eloquently than any man can describe it the story of the abundance of game in that easily accessible wilderness. "Dat red deer," Antoine replied. "Him fox." "Dat bear track; dat squirrel; dat rabbit." "Dat moose track; pass las' week." "Dat pa'tridge; dat wolf." Or perhaps it was the trail of a marten, or a beaver, or a weasel, or a fisher, mink, lynx, or otter that he pointed out, for all these "signs" were there, and nearly all were repeated again and again. Of the birds that are plentiful there the principal kinds are partridge, woodcock, crane, geese, duck, gull, loon, and owl.

When the sun set we prepared to camp, selecting a spot near a tiny rill. The horses were tethered to a tree, with their harness still on, and blankets thrown over them. We cleared a little space by the roadside, using our snow-shoes for shovels. The Indians, with their axes, turned up the moss and leaves, and levelled the small shoots and brushwood. Then one went off to cut balsam boughs for bedding, while the other set up two crotched sticks, with a pole upon them resting in the crotches, and throwing the canvas of an "A" tent over the frame, he looped the bottom of the tent to small pegs, and banked snow lightly all around it. The little aromatic branches of balsam were laid evenly upon the ground, a fur robe was thrown upon the leaves, our enormous blankets were spread half open side by side, and two coats were rolled up and thrown down for pillows. Pierre, the second Indian, made tiny slivers of some soft wood, and tried to start a fire. He failed. Then Alexandre Antoine brought two handfuls of bark, and lighting a small piece with a match, proceeded to build a fire in the most painstaking manner, and with an ingenuity that was most interesting. First he made a fire that could have been started in a teacup; then he built above and around it a skeleton tent of bits of soft wood, six to nine inches in length. This gave him a fire of the dimensions of a high hat. Next, he threw down two great bits of timber, one on either side of the fire, and a still larger back log,

THE TRACK IN THE WINTER FOREST

and upon these he heaped split soft wood. While this was being done, Pierre assailed one great tree after another, and brought them crashing down with noises that startled the forest quiet. Alphonse had opened the provision bags, and presently two tin pails filled with water swung from saplings over the fire, and a pan of fat salt pork was frizzling upon the blazing wood. The darkness grew dead black, and the dancing flames peopled the near forest with dodging shadows. Almost in the time it has taken me to write it, we were squatting on our heels around the fire, each with a massive cutting of bread, a slice of fried pork in a tin plate, and half a pint of tea, precisely as hot as molten lead, in a tin cup. Supper was a necessity, not a luxury, and was hurried out of the way accordingly. Then the men built their camp beside ours in front of the fire, and followed that by felling three or more monarchs of the bush. Nothing surprised me so much as the amount of wood consumed in these open-air fires. In five days at our permanent camp we made a great hole in the forest.

But that first night in the open air, abed with nature, with British America for a bedroom! Only I can tell of it, for the others slept. The stillness was intense. There was no wind, and not an animal or bird uttered a cry. The logs cracked and sputtered and popped, the horses shook their chains, the men all snored—white and red alike. The horses pounded the hollow earth; the logs broke and fell upon the cinders; one of the men talked in his sleep. But over and through it all the stillness grew. Then the fire sank low, the cold became intense, the light was lost, and the darkness swallowed everything. Some one got up awkwardly, with muttering, and flung wood upon the red ashes, and presently all that had passed was re-experienced.

The ride next day was more exciting than the first stage. It was like the journey of a gun-carriage across country in a hot retreat. The sled was actually upset only once, but to prevent that happening fifty times the Indians kept springing at the uppermost side of the flying vehicle, and hanging to the side poles to pull the toppling construction down upon both runners. Often we were advised to leap out for safety's sake; at other times we wished we had leaped out. For seven hours we were flung about like cotton spools that are being polished in a revolving cylinder. And yet we were obliged to run long distances after the hurtling sleigh — long enough to tire us. The artist, who had spent years in rude scenes among rough men, said nothing at the time. What was the use? But afterwards, in New York, he remarked that this was the roughest travelling he had ever experienced.

The signs of game increased. Deer and bear and wolf and fox and moose were evidently numerous around us. Once we stopped, and the Indians became excited. What they had taken for old moose tracks were the week-old footprints of a man. It seems strange, but they felt obliged to know what a man had gone into the bush for a week ago. They followed the signs, and came back smiling. He had gone in to cut hemlock boughs; we would find traces of a camp near by. We did. In a country where men are so few, they busy themselves about one another. Four or five days later, while we were hunting, these Indians came to the road and stopped suddenly, as horses do when lassoed. With a glance they read that two teams had passed during the night, going towards our camp. When we returned to camp the teams had been there, and our teamster had talked with the drivers. Therefore

PIERRE, FROM LIFE.

that load was lifted from the minds of our Indians. But their knowledge of the bush was marvellous. One point in the woods was precisely like another to us, yet the Indians would leap off the sleigh now and then and dive into the forest, to return with a trap hidden there months before, or to find a great iron kettle.

"Do you never get lost?" I asked Alexandre.

"Me get los'? No, no get los'."

"But how do you find your way?"

"Me fin' way easy. Me know way me come, or me follow my tracks, or me know by de sun. If no sun, me look at trees. Trees grow more branches on side toward sun, and got rough bark on north side. At night me know by see de stars."

We camped in a log-hut Alexandre had built for a hunting camp. It was very picturesque and substantial, built of huge logs, and caulked with moss. It had a great earthen bank in the middle for a fireplace, with an equally large opening in the roof, boarded several feet high at the sides to form a chimney. At one corner of the fire bank was an ingenious crane, capable of being raised and lowered, and projecting from a pivoted post, so that the long arm could be swung over or away from the fire. At one end of the single apartment were two roomy bunks built against the wall. With extraordinary skill and quickness the Indians whittled a spade out of a board, performing the task with an axe, an implement they can use as white men use a penknife, an implement they value more highly than a gun. They made a broom of balsam boughs, and dug and swept the dirt off the floor and walls, speedily making the cabin neat and clean. Two new bunks were put up for us, and bedded with balsam boughs and skins. Shelves were already up, and spread with pails and bottles, tin cups and plates, knives and forks, canned goods, etc. On them and on the floor were our stores.

We had a week's outfit, and we needed it, because for five days we could not hunt on account of the crust on the snow, which made such a noise when a human foot broke through it that we could not have approached any wild animal within half a mile. On the third day it rained, but without melting the

ANTOINE'S CABIN

crust. On the fourth day it snowed furiously, burying the crust under two inches of snow. On the fifth day we got our moose.

In the mean time the log-cabin was our home. Alexandre and Pierre cut down trees every day for the fire, and Pierre disappeared for hours every now and then to look after traps set for otter, beaver, and marten. Alphonse attended his horses and served as cook. He could produce hotter tea than any other man in the world. I took mine for a walk in the arctic cold three times a day, the artist learned to pour his from one cup to another with amazing dexterity, and the Indians (who drank a quart each of green tea at each meal because it was stronger than our black tea) lifted their pans and threw the liquid fire down throats that had been inured to high wines. Whenever the fire was low, the cold was intense. Whenever it was heaped with logs, all the heat flew directly through the roof, and spiral blasts of cold air were sucked through every crack between logs in the cabin walls. Whenever the door opened, the cabin filled with smoke. Smoke clung to all we ate or wore. At night the fire kept burning out, and we arose with chattering teeth to build it anew. The Indians were then to be seen with their blankets pushed down to their knees, asleep in their shirts and trousers. At meal-times we had bacon or pork, speckled or lake trout, bread-and-butter, stewed tomatoes, and tea. There were two stools for the five men, but they only complicated the discomfort of those who got them; for it was found that if we put our tin plates on our knees, they fell off; if we held them in one hand, we could not cut the pork and hold the bread with the other hand; while if we put the plates on the floor beside the tea, we could not reach them. In a month we might have solved the problem. Life in that log shanty was precisely the life of the early settlers of this country. It was bound to produce great characters or early death. There could be no middle course with such an existence.

Partridge fed in the brush impudently before us. Rabbits bobbed about in the clearing before the door. Squirrels sat upon the logs near by and gormandized and chattered. Great saucy birds, like mouse-colored robins, and known to the Indians as "meat-birds," stole our provender if we left it out-of-doors half an hour, and one day we saw a red deer jump in the bush a hundred yards away. Yet we got no game, because we knew there was a moose-yard within two miles on one side and within three miles on the other, and we dared not shoot our rifles lest we frighten the moose. Moose was all we were after. There was a lake near by, and the trout in those lakes up there attain remarkable size and numbers. We heard of 35-pound speckled trout, of lake trout twice as large, and of enormous muskallonge. The most reliable persons told of lakes farther in the wilderness where the trout are thick as salmon in the British Columbia streams—so thick as to seem to fill the water. We were near a lake that was supposed to have been fished out by lumbermen a year before, yet it was no sport at all to fish there. With a short stick and two yards of line and a bass

THE CAMP AT NIGHT

hook baited with pork, we brought up four-pound and five-pound beauties faster than we wanted them for food. Truly we were in a splendid hunting country, like the Adirondacks eighty years ago, but thousands of times as extensive.

Finally we started for moose. Our Indians asked if they might take their guns. We gave the permission. Alexandre, a thin, wiry man of forty years, carried an old Henry rifle in a woollen case open at one end like a stocking. He wore a short blanket coat and tuque, and trousers tied tight below the knee, and let into his moccasin-tops. He and his brother François are famous Hudson Bay Company trappers, and are two-thirds Algonquin and one-third French. He has a typical swarthy, angular Indian face and a French mustache and goatee. Naturally, if not by rank, a leader among his men, his manner is commanding and his appearance grave. He talks bad French fluently, and makes wretched headway in English. Pierre is a short, thickset, walnut-stained man of thirty-five, almost pure Indian, and almost a perfect specimen of physical development. He seldom spoke while on this trip, but he impressed us with his strength, endurance, quickness, and knowledge of woodcraft. Poor fellow! he had only a shotgun, which he loaded with buckshot. It had no case, and both men carried their pieces grasped by the barrels and shouldered, with the butts behind them.

We set out in Indian-file, plunging at once into the bush. Never was forest scenery more exquisitely beautiful than on that morning as the day broke, for we breakfasted at four o'clock, and started immediately afterwards. Everywhere the view was fairy-like. There was not snow enough for snowshoeing. But the fresh fall of snow was immaculately white, and flecked the scene apparently from earth to sky, for there was not a branch or twig or limb or spray of evergreen, or wart or fungous growth upon any tree that did not bear its separate burden of snow. It was a bridal dress, not a winding-sheet, that Dame Nature was trying on that morning. And in the bright fresh green of the firs and pines we saw her complexion peeping out above her spotless gown, as one sees the rosy cheeks or black eyes of a girl wrapped in ermine.

Mile after mile we walked, up mountain and down dale, slapped in the faces by twigs, knocking snow down the backs of our necks, slipping knee-deep in bog mud, tumbling over loose stones, climbing across interlaced logs, dropping to the height of one thigh between tree trunks, sliding, falling, tight-rope walking on branches over thin ice, but forever following the cat-like tread of Alexandre, with his seven-league stride and long-winded persistence. Suddenly we came to a queer sort of clearing dotted with protuberances like the bubbles on molasses beginning to boil. It was a beaver meadow. The bumps in the snow covered stumps of trees the beavers had gnawed down. The Indians were looking at some trough like tracks in the snow, like the trail of a tired man who had dragged his heels. "Moose; going this way," said Alexandre; and we turned and walked in the tracks. Across the meadow and across a lake

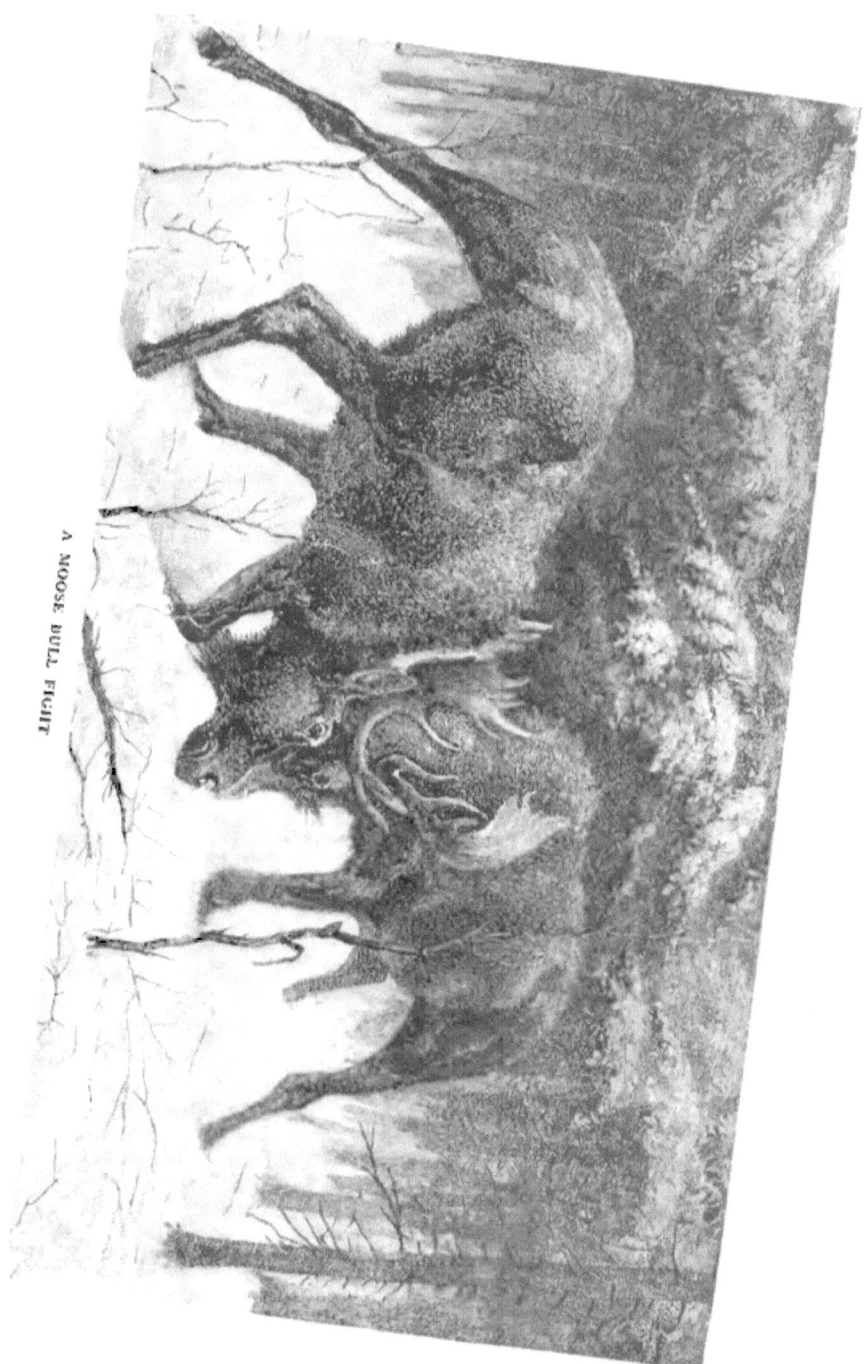

A MOOSE BULL FIGHT

and up another mountain they led us. Then we came upon fresher prints. At each new track the Indians stooped, and making a scoop of one hand, brushed the new-fallen snow lightly out of the indentations. Thus they read the time at which the print was made. "Las' week," "Day 'fore yesterday," they whispered. Presently they bent over again, the light snow flew, and one whispered, "This morning."

Stealthily Alexandre swept ahead; very carefully we followed. We dared not break a twig, or speak, or slip, or stumble. As it was, the breaking of the crust was still far too audible. We followed a little stream, and approached a thick growth of tamarack. We had no means of knowing that a herd of moose was lying in that thicket, resting after feeding. We knew it afterwards. Alexandre motioned to us to

ON THE MOOSE TRAIL

get our guns ready. We each threw a cartridge from the cylinder into the barrel, making a "click, click" that was abominably loud. Alexandre forged ahead. In five minutes we heard him call aloud: "Moose gone. We los' him." We hastened to his side. He pointed at some tracks in which the prints were closer together than any we had seen.

"See! he trot," Alexandre explained.

In another five minutes we had all but completed a circle, and were on the other side of the tamarack thicket. And there were the prints of the bodies of the great beasts. We could see even the imprint of the hair of their coats. All around were broken twigs and balsam needles. The moose had left the branches ragged, and on every hand the young bark was chewed or rubbed raw. Loading our rifles had lost us a herd of moose.

Back once again at the beaver dam, Alexandre and Pierre studied the moose-tramped snow and talked earnestly. They agreed that a desperate battle had been fought there between two bull moose a week before, and that those bulls were not in the "yard" where we had blundered. They examined the tracks over an acre or more, and then strode off at an obtuse angle from our former trail. Pierre, apparently not quite satisfied, kept dropping behind or disappearing in the bush at one side of us. So magnificent was his skill at his work that I missed him at times, and at other times found him putting his feet down where mine were lifted up without ever hearing a sound of his step or of his contact with the undergrowth. Alexandre presently motioned us

IN SIGHT OF THE GAME—"NOW SHOOT!"

with a warning gesture. He slowed his pace to
short steps, with long pauses between. He saw
everything that moved, heard every sound; only a
deer could throw more and keener faculties into play
than this born hunter. He heard a twig snap. We
heard nothing. Pierre was away on a side search.
Alexandre motioned us to be ready. We crept close
together, and I scarcely breathed. We moved cau-
tiously, a step at a time, like chessmen. It was impossible to get an unobstructed view a hundred feet
ahead, so thick was the soft-wood growth. It seemed
out of the question to try to shoot that distance.
We were descending a hill-side into marshy ground.
We crossed a corner of a grove of young alders, and
saw before us a gentle slope thickly grown with
evergreen—tamarack, the artist called it. Suddenly
Alexandre bent forward and raised his gun. Two
steps forward gave us his view. Five moose were
fifty yards away, alarmed and ready to run. A big
bull in the front of the group had already thrown
back his antlers. By impulse rather than through
reason I took aim at a second bull. He was half a
height lower down the slope, and to be seen through
a web of thin foliage. Alexandre and the artist fired
as with a single pull at one trigger. The foremost
bull staggered and fell forward, as if his knees had
been broken. He was hit twice—in the heart and
in the neck. The second bull and two cows and a
calf plunged into the bush and disappeared. Pierre
found that bull a mile away, shot through the lungs.

It had taken us a week to kill our moose in a
country where they were common game. That was

"hunter's luck" with a vengeance. But at another season such a delay could scarcely occur. The time to visit that district is in the autumn, before snow falls. Then in a week one ought to be able to bag a moose, and move into the region where caribou are plenty.

Mr. Remington, in the picture called "Hunting the Caribou," depicts a scene at a critical moment in the experience of any man who has journeyed on westward of where we found our moose, to hunt the caribou. There is a precise moment for shooting in the chase of all animals of the deer kind, and when that moment has been allowed to pass, the chance of securing the animal diminishes with astonishing rapidity—with more than the rapidity with which the then startled animal is making his flight, because to his flight you must add the increasing ambush of the forest. What is true of caribou in this respect is true of moose and red deer, elk and musk-ox in America, and of all the horned animals of the forests of the other great hemisphere. Every hunter who sees Mr. Remington's realistic picture knows at a glance that the two men have stolen noiselessly to within easy rifle-shot of a caribou, and that suddenly, at the last moment, the animal has heard them.

Perhaps he has seen them, and is standing—still as a Barye bronze—with his great, soft, wondering eyes riveted upon theirs. That is a situation familiar to every hunter. His prey has been browsing in fancied security, and yet with that nervous prudence that causes these timid beasts to keep forever raising their heads, and sweeping the view around them

SUCCESS

with their exquisite sight, and analyzing the atmosphere with their magical sense of smell. In one of these cautious pauses the caribou has seen the hunters. Both hunters and hunted seem instantly to turn to stone. Neither moves a muscle or a hair. If the knee or the foot of one of the men presses too hard upon a twig and it snaps, the caribou is as certain to throw his head high up and dart into the ingulfing net-work of the forest trunks and brush as day is certain to follow night. But when no movement has been made and no mishap has alarmed the beast, it has often happened that the two or more parties to this strangely thrilling situation have held their places for minutes at a stretch — minutes that seemed like quarters of an hour. In such cases the deer or caribou has been known to lower his head and feed again, assured in its mind that the suspected hunter is inanimate and harmless. Nine times in ten, though, the first to move is the beast, which tosses up its head, and "Shoot! shoot!" is the instant command, for the upward throwing of the head is a movement made to put the beast's great antlers into position for flight through the forest.

The caribou has very wide, heavy horns, and they are almost always circular—that is, the main part or trunk of each horn curves outward from the skull and then inward towards the point, in an almost true semicircle. They are more or less branched, but both the general shape of the whole horns and of the branches is such that when the head is thrown up and back they aid the animal's flight by presenting what may be called the point of a wedge towards the saplings and limbs and small forest growths through which the beast runs, parting and spreading every pair of obstacles to either side, and bending every single one out of the way of his flying body. The caribou of North America is the reindeer of Greenland; the differences between the two are very slight. The animal's home is the arctic circle, but in America it feeds and roams farther south than in Europe and Asia. It is a large and clumsy-looking beast, with thick and rather short legs and bulky body, and, seen in repose, gives no hint of its capacity for flight. Yet the caribou can run "like a streak of wind," and makes its way through leaves and brush and brittle, sapless vegetation with a modicum of noise so slight as to seem inexplicable. Nature has ingeniously added to its armament, always one, and usually two, palmated spurs at the root of its horns, and these

grow at an obtuse angle with the head, upward and outward towards the nose. With these spurs—like shovels used sideways—the caribou roots up the snow, or breaks its crust and disperses it, to get at his food on the ground. The caribou are very large deer, and their strength is attested by the weight of their horns. I have handled caribou horns in Canada that I could not hold out with both hands when seated in a chair. It seemed hard to believe that an animal of the size of a caribou could carry a burden apparently so disproportioned to his head and neck. But it is still more difficult to believe, as all the woodsmen say, that these horns are dropped and new ones grown every year.

It is not the especial beauty of Frederic Remington's drawings and paintings that they are absolutely accurate in every detail, but it is one of their beauties, and gives them especial value apart from their artistic excellence. He draws what he knows, and he knows what he draws. This scene of the electrically exquisite moment in a hunter's life, when great game is before him, and the instant has come for claiming it as his own with a steadily held and wisely chosen aim, will give the reader a perfect knowledge of how the Indians and hunters dress and equip themselves beyond the Canadian border. The scene is in the wilderness north of the Great Lakes. The Indian is of one of those tribes that are offshoots of the great Algonquin nation. He carries in that load he bears that which the plainsmen call " the grub stake," or quota of provisions for himself and his employer, as well as blankets to sleep in, pots, pans, sugar, the inevitable tea of those latitudes, and much else besides. Those Indians are not as lazy or as physically degenerate as many of the tribes in our country. They turn themselves into wonderful beasts of burden, and go forever equipped with a long, broad strap that they call a "tomp line," and which they pass around their foreheads and around their packs, the latter resting high up on their backs. It seems incredible, but they can carry one hundred to one hundred and fifty pounds of necessaries all day long in the roughest regions. The Hudson Bay Company made their ancestors its wards and dependents two centuries ago, and taught them to work and to earn their livelihood.

BIG FISHING

IN October every year there are apt to be more fish upon the land in the Nepigon country than one would suppose could find life in the waters. Most families have laid in their full winter supply, the main exceptions being those semi-savage families which leave their fish out — in preference to laying them in — upon racks whereon they are to be seen in rows and by the thousands.

Nepigon, the old Hudson Bay post which is the outfitting place for this region, is 928 miles west of Montreal, on the Canadian Pacific Railway, and on an arm of Lake Superior. The Nepigon River, which connects the greatest of lakes with Lake Nepigon, is the only roadway in all that country, and therefore its mouth, in an arm of the great lake, is the front door to that wonderful region. In travelling through British Columbia I found one district that is going to prove of greater interest to gentlemen sportsmen with the rod, but I know of no greater fishing country than the Nepigon. No single waterway or system of navigable inland waters in North America is likely to wrest the palm from this Nepigon district as the haunt of fish in the greatest plenty, unless we term the salmon a fresh-water fish, and thus call the Fraser, Columbia, and Skeena rivers into the rivalry. There is incessant fishing in this wilderness north of Lake Superior from New-year's Day, when the ice has to be cut to get at the water, all through the succeeding seasons, until again the ice fails to protect the game. And there is every sort of fishing between that which engages a navy of sailing vessels and men, down through all the methods of fish-taking — by nets, by spearing, still fishing, and fly-fishing. A half a dozen sorts of finny game succumb to these methods, and though the region has been famous and therefore much visited for nearly a dozen years, the field is so extensive, so well stocked, and so difficult of access except to persons of means, that even to-day almost the very largest known specimens of each class of fish are to be had there.

If we could put on wings early in October, and could fly down from James's Bay over the dense forests and countless lakes and streams of western Ontario, we would see now and then an Indian or hunter in a canoe, here and there a lonely huddle of small houses forming a Hudson Bay post, and at even greater distances apart small bunches of the cotton or birch-bark tepees of pitiful little Cree or Ojibaway bands. But with the first glance at the majestic expanse of Lake Superior there would burst upon the view scores upon scores of white sails upon the water, and near by, upon the shore, a tent for nearly every sail. That is the time for the annual gathering for catching the big, chunky, red-fleshed fish they call the salmon-trout. They catch those that weigh from a dozen to twenty-five or thirty pounds, and at this time of the year their flesh is comparatively hard.

Engaged in making this great catch are the boats of the Indians from far up the Nepigon and the neighboring streams; of the chance white men of the region, who depend upon nature for their sustenance; and of Finns, Norwegians, Swedes, and others who come from the United States side, or southern shore, to fish for their home markets. These fish come at this season to spawn, seeking the reefs, which are plentiful off the shore in this part of the lake. Gill nets are used to catch them, and are set within five fathoms of the surface by setting the inner buoy in water of that depth, and then paying the net out into deeper water and anchoring it. The run and the fishing continue throughout October. As a rule, among the Canadians and Canada Indians a family goes with each boat—the boats being sloops of twenty-seven to thirty feet in length, and capable of carrying fifteen pork barrels, which are at the outset filled with rock-salt. Sometimes the heads of two families are partners in the ownership of one of these sloops, but, however that may be, the custom is for the women and children to camp in tents along-shore, while the men (usually two men and a boy for each boat) work the nets. It is a stormy season of the year, and the work is rough and hazardous, especially for the nets, which are frequently lost.

Whenever a haul is made the fish are split down the back and cleaned. Then they are washed, rolled in salt, and packed in the barrels. Three days later, when the bodies of the fish have thoroughly purged themselves, they are taken out, washed again, and are once more rolled in fresh salt and put back in the barrels, which are then filled to the top with water. The Indians subsist all winter upon this October catch, and, in addition, manage to exchange a few barrels for other provisions and for clothing. They demand an equivalent of six dollars a barrel in whatever they get in exchange, but do not sell for money, because, as I understand it, they are not obliged to pay the provincial license fee as fishermen, and therefore may not fish for the market. Even sportsmen who throw a fly for one day in the Nepigon country must pay the Government for the privilege. The Indians told me that eight barrels of these fish will last a family of six persons an entire winter. Such a demonstration of prudence and forethought as this, of a month's fishing at the threshold of winter, amounts to is a rare one for an Indian to make, and I imagine there is a strong admixture of white blood in most of those who make it. The full-bloods will not take the trouble. They trust to their guns and their traps against the coming of that wolf which they are not unused to facing.

Up along the shores of Lake Nepigon, which is thirty miles by an air line north of Lake Superior, many of the Indians lay up white-fish for winter. They catch them in nets and cure them by frost. They do not clean them. They simply make a hole in the tail end of each fish, and string them, as if they were beads, upon sticks, which they set up into racks. They usually hang the fishes in rows

INDIANS HAULING NETS ON LAKE NEPIGON

of ten, and frequently store up thousands while they are at it. The Reverend Mr. Renison, who has had much to do with bettering the condition of these Indians, told me that he had caught 1020 pounds of white-fish in two nights with two gill nets in Lake Nepigon. It is unnecessary to add that he cleaned his.

Lake Nepigon is about seventy miles in length, and two-thirds as wide, at the points of its greatest measurement, and is a picturesque body of water, surrounded by forests and dotted with islands. It is a famous haunt for trout, and those fishermen who are lucky may at times see scores of great beauties lying upon the bottom; or, with a good guide and at the right season, may be taken to places where the water is fairly astir with them. Fishermen who are not lucky may get their customary experience without travelling so far, for the route is by canoe, on top of nearly a thousand miles of railroading; and one mode of locomotion consumes nearly as much time as the other, despite the difference between the respective distances travelled. The speckled trout in the lake are locally reported to weigh from three to nine pounds, but the average stranger will lift in more of three pounds' weight than he will of nine. Yet whatever they average, the catching of them is prime sport as you float upon the water in your picturesque birch-bark canoe, with your guide paddling you noiselessly along, and your spoon or artificial minnow rippling through the water or glinting in the sunlight. You need a stout bait-rod, for the gluttonous fish are game, and make a good fight every time. The local fishermen catch the speckled beauties with an unpoetic lump of pork.

A lively French Canadian whom I met on the cars on my way to Nepigon described that region as "de mos' tareeble place for de fish in all over de worl'." And he added another remark which had at least the same amount of truth at the bottom of it. Said he: "You weel find dere dose Mees Nancy feeshermans from der Unite State, vhich got dose hunderd-dollar poles and dose leetle humbug flies, vhich dey t'row around and pull 'em back again, like dey was afraid some feesh would bite it. Dat is all one grand stupeedity. Dose man vhich belong dere put on de hook some pork, and catch one tareeble pile of fish. Dey don't give a —— about style, only to catch dose feesh."

To be sure, every fisherman who prides himself on the distance he can cast, and who owns a splendid outfit, will despise the spirit of that French Canadian's speech; yet up in that country many a scientific angler has endured a failure of "bites" for a long and weary time, while his guide was hauling in fish a-plenty, and has come to question "science" for the nonce, and follow the Indian custom. For gray trout (the namaycush, or lake trout) they bait with apparently anything edible that is handiest, preferring pork, rabbit, partridge, the meat of the trout itself, or of the sucker; and the last they take first, if possible. The suckers, by-the-way, are all too plenty, and as full of bones as any old-time frigate ever was with timbers. You may see the Indians eating them and discarding the bones at the same time; and they make the process resemble the action of a hay-cutter when the grass is going in long at one side, and coming out short, but in equal quantities, at the other.

The namaycush of Nepigon weigh from nine to twenty-five pounds. The natives take a big hook and bait it, and then run the point into a piece of shiny, newly-scraped lead. They never "play" their bites, but give them a tight line and steady pull. These fish make a game struggle, leaping and diving and thrashing the water until the gaff ends the struggle. In winter there is as good sport with the namaycush, and it is managed peculiarly. The Indians cut into the ice over deep water, making holes at least eighteen inches in diameter. Across the hole they lay a stick, so that when they pull up a trout the line will run along the stick, and the fish will hit that obstruction instead of the resistant ice. If a fish struck the ice the chances are nine to one that it would tear off the hook. Having baited a hook with pork, and stuck the customary bit of lead upon it, they sound for bottom, and then measure the line so that it will reach to about a foot and a half above soundings—that is to say, off bottom. Then they begin fishing, and their plan is (it is the same all over the Canadian wilderness) to keep jerking the line up with a single, quick, sudden bob at frequent intervals.

The spring is the time to catch the big Nepigon jack-fish, or pike. They haunt the grassy places in little bogs and coves, and are caught by trolling. A jack-fish is what we call a pike, and John Watt, the famous guide in that country, tells of those fish of such size that when a man of ordinary height held the tail of one up to his shoulder, the head of the fish dragged on the ground. He must be responsible for the further assertion that he saw an Indian squaw drag a net, with meshes seven inches square, and catch two jack-fish, each of which weighed more than fifty pounds when cleaned. The story another local historian told of a surveyor who caught a big jack-fish that felt like a sunken log, and could only be dragged until its head came to the surface, when he shot it and it broke away—that narrative I will leave for the next New Yorker who goes to Nepigon. And yet it seems to me that such stories distinguish a fishing resort quite as much as the fish actually caught there. Men would not dare to romance like that at many places I have fished in, where the trout are scheduled and numbered, and where you have got to go to a certain rock on a fixed day of the month to catch one.

The Indians are very clever at spearing the jack-fish. At night they use a bark torch, and slaughter the big fish with comparative ease; but their great skill with the spear is shown in the daytime, when the pike are sunning themselves in the grass and weeds along-shore. But when I made my trip up the river, I saw them using so many nets as to threaten the early reduction of the stream to the plane of the ordinary resort. The water was so clear that we could paddle beside the nets and see each one's catch—here a half-dozen suckers, there a jack-fish, and next a couple of beautiful trout. Finding a squaw attending to her net, we bought a trout from her before we had cast a line. The habit of buying fish under such circumstances becomes second nature to a New Yorker. We are a peculiar people. Our fishermen are modest away from the city, but at home they assume the confident tone which comes of knowing the way to Fulton fish-market.

The Nepigon River is a trout's paradise, it is so full of rapids and saults. It is not at all a folly to fish there with a fly-rod. There are records of very large trout at the Hudson Bay post; but you may actually catch four-pound trout yourself, and what you catch yourself seems to me better than any one's else records. I have spoken of the Nepigon River as a roadway. It is one of the great trading trails to and from the far North. At the mouth of the river, opposite the Hudson Bay post, you will see a wreck of one of its noblest vehicles—an old York boat, such as carry the furs and the supplies to and fro. I fancy that Wolseley used precisely such boats to float his men to where he wanted them in 1870. Farther along, before you reach the first portage, you will be apt to see several of the sloops used by the natives for the Lake Superior fishing. They are distinguished for their ugliness, capacity, and strength; but the last two qualities are what they are built to obtain. Of course the prettiest vehicles are the canoes. As the bark and the labor are easily obtainable, these picturesque vessels are very numerous; but a change is coming over their shape, and the historic Ojibaway canoe, in which Hiawatha is supposed to have sailed into eternity, will soon be a thing found only in pictures.

There is good sport with the rod wherever you please to go in "the bush," or wilderness, north of the Canadian Pacific Railway, in Ontario and the western part of Quebec. My first venture in fishing through the ice in that region was part of a hunting experience, when the conditions were such that hunting was out of the question, and our party feasted upon salt pork, tea, and tomatoes during day after day. At first, fried salt pork, taken three times a day in a hunter's camp, seems not to deserve the harsh things that have been said and written about it. The openair life, the constant and tremendous exercise of hunting or chopping wood for the fire, the novel surroundings in the forest or the camp, all tend to make a man say as hearty a grace over salt pork as he ever did at home before a holiday dinner. Where we were, up the Ottawa in the Canadian wilderness, the pork was all fat, like whale blubber. At night the cook used to tilt up a pan of it, and put some twisted ravellings of a towel in it, and light one end, and thus produce a lamp that would have turned Alfred the Great green with envy, besides smoking his palace till it looked as venerable as Westminster Abbey does now. I ate my share seasoned with the comments of Mr. Frederic Remington, the artist, who asserted that he was never without it on his hunting trips, that it was pure carbonaceous food, that it fastened itself to one's ribs like a true friend, and that no man could freeze to death in the same country with this astonishing provender. We had canned tomatoes and baker's bread and plenty of tea, with salt pork as the *pièce de résistance* at every meal. I know now—though I would not have confessed it at the time—that mixed with my admiration of salt pork was a growing dread that in time, if no change offered itself, I should tire of that diet. I began to feel it sticking to me more like an Old Man of the Sea than a brother. The woodland atmosphere began to taste of it. When I came in-doors it seemed to me that the log shanty was gradually turning into fried salt pork. I could not say that I knew how it felt to eat a quail a day for thirty days. One man cannot know everything. But I felt that I was learning.

One day the cook put his hat on, and took his axe, and started out of the shanty door with an unwonted air of business.

"Been goin' fish," said he, in broken Indian. "Good job if get trout."

A good job? Why, the thought was like a floating spar to a sailor overboard! I went with him. It was a cold day, but I was dressed in Canadian style—the style of a country where every one puts on everything he owns: all his stockings at once, all his flannel shirts and drawers, all his coats on top of one another, and when there is nothing else left, draws over it all a blanket suit, a pair of moccasins, a tuque, and whatever pairs of gloves he happens to be able to find or borrow. One gets a queer feeling with so many clothes on. They seem to separate you from yourself, and the person you feel inside your clothing might easily be mistaken for another individual. But you are warm, and that's the main thing.

I rolled along the trail behind the Indian, through the deathly stillness of the snow-choked forest, and presently, from a knoll and through an opening, we saw a great woodland lake. As it lay beneath its unspotted quilt of snow, edged all around with balsam, and pine and other evergreens, it looked as though some mighty hand had squeezed a colossal tube of white paint into a tremendous emerald bowl. Never had I seen nature so perfectly unalloyed, so exquisitely pure and peaceful, so irresistibly beauti-

ful. I think I should have hesitated to print my ham-like moccasin upon that virgin sheet had I been the guide, but "Brossy," the cook, stalked ahead, making the powdery flakes fly before and behind him, and I followed. Our tracks were white, and quickly faded from view behind us; and, moreover, we passed the signs of a fox and a deer that had crossed during the night, so that our profanation of the scene was neither serious nor exclusive.

The Indian walked to an island near the farther shore, and using his axe with the light, easy freedom that a white man sometimes attains with a penknife, he cut two short sticks for fish-poles. He cut six yards of fish-line in two in the middle of the piece, and tied one end of each part to one end of each stick, making rude knots, as if any sort of a fastening would do. Equally clumsily he tied a bass hook to each fish-line, and on each hook he speared a little cube of pork fat which had gathered an envelope of granulated smoking-tobacco while at rest in his pocket. Next, he cut two holes in the ice, which was a foot thick, and over these we stood, sticks in hand, with the lines dangling through the holes. Hardly had I lowered my line (which had a bullet flattened around it for a sinker, by-the-way) when I felt it jerked to one side, and I pulled up a three-pound trout. It was a speckled trout. This surprised me, for I had no idea of catching anything but lake or gray trout in that water. I caught a gray trout next—a smaller one than the first—and in another minute I had landed another three-pound speckled beauty. My pork bait was still intact, and it may be of interest to fishermen to know that the original cubes of pork remained on those two hooks a week, and caught us many a mess of trout.

There came a lull, which gave us time to philosophize on the contrast between this sort of fishing and the fashionable sport of using the most costly and delicate rods—like pieces of jewelry—and of calculating to a nicety what sort of flies to use in matching the changing weather or the varying tastes of trout in waters where even all these calculations and provisions would not yield a hatful of small fish in a day. Here I was, armed like an urchin beside a minnow brook, and catching bigger trout than I ever saw outside Fulton Market—trout of the choicest variety. But while I moralized my Indian grew impatient, and cut himself a new hole out over deep water. He caught a couple of two-and-a-half-pound brook trout and a four-pound gray trout, and I was as well rewarded. But he was still discontented, and moved to a strait opening into a little bay, where he cut two more holes. "Eas' wind," said he, "fish no bite."

I found on that occasion that no quantity of clothing will keep a man warm in that almost arctic climate. First my hands became cold, and then my feet, and then my ears. A thin film of ice closed up the fishing holes if the water was not constantly disturbed. The thermometer must have registered ten or fifteen degrees below zero. Our lines became quadrupled in thickness at the lower ends by the ice that formed upon them. When they coiled for an instant upon the ice at the edge of a hole, they stuck to it, frozen fast. By stamping my feet and putting my free hand in my pocket as fast as I shifted my pole from one hand to the other, I managed to persist in fishing. I noticed many interesting things as I stood there, almost alone in that almost pathless wilderness. First I saw that the Indian was not cold, though not half so warmly dressed as I. The circulation or vitality of those scions of nature must be very remarkable, for no sort of weather seemed to trouble them at all. Wet feet, wet bodies, intense cold, whatever came, found and left them indifferent. Night after night, in camp, in the open air, or in our log shanty, we white men trembled with the cold when the log fire burned low, but the Indians never woke to rebuild it. Indeed, I did not see one have his blanket pulled over his chest at any time. Woodcocks were drumming in the forest now and then, and the shrill, bird-like chatter of the squirrels frequently rang out upon the forest quiet. My Indian knew every noise, no matter how faint, yet never raised his head to listen. " Dat squirrel," he would say, when I asked him. Or, " Woodcock, him calling rain," he ventured. Once I asked what a very queer, distant, muffled sound was. " You hear dat when you walk. Keep still, no hear dat," he said. It was the noise the ice made when I moved.

As I stood there a squirrel came down upon a log jutting out over the edge of the lake, and looked me over. A white weasel ran about in the bushes so close to me that I could have hit him with a peanut shell. That morning some partridge had been seen feeding in the bush close to members of our party.

It was a country where small game is not hunted, and does not always hide at man's approach. We had left our fish lying on the ice near the various holes from which we pulled them, and I thought of them when a flock of ravens passed overhead, crying out in their hoarse tones. They were sure to see the fish dotting the snow like raisins in a bowl of rice.

"Won't they steal the fish?" I asked.

"T'ink not," said the Indian.

"I don't know anything about ravens," I said, "but if they are even distantly related to a crow, they will steal whatever they can lift."

We could not see our fish around the bend of the lake, so the Indian dropped his rod and walked stolidly after the birds. As soon as he passed out of sight I heard him scolding the great birds as if they were unruly children.

"'Way, there!" he cried—"'way! Leave dat fish, you. What you do dere, you t'ief?"

It was an outcropping of the French blood in his veins that made it possible for him to do such violence to Indian reticence. The birds had seen our fish, and were about to seize them. Only the foolish bird tradition that renders it necessary for everything with wings to circle precisely so many times over its prey before taking it saved us our game and lost them their dinner. They had not completed half their quota of circles when Brossy began to yell at them. When he returned his brain had awakened, and he began to remember that ravens were thieves. He said that the lumbermen in that country pack their dinners in canvas sacks and hide them in the snow. Often the ravens come, and, searching out this food, tear off the sacks and steal their contents. I bade good-bye to pork three times a day after that. At least twice a day we feasted upon trout.

VI

"A SKIN FOR A SKIN"

The motto of the Hudson Bay Fur-trading Company

THOSE who go to the newer parts of Canada today will find that several of those places which their school geographies displayed as Hudson Bay posts a few years ago are now towns and cities. In them they will find the trading stations of old now transformed into general stores. Alongside of the Canadian headquarters of the great corporation, where used to stand the walls of Fort Garry, they will see the principal store of the city of Winnipeg, an institution worthy of any city, and more nearly to be likened to Whiteley's Necessary Store in London than to any shopping-place in New York. As in Whiteley's you may buy a house, or anything belonging in or around a house, so you may in this great Manitoban establishment. The great retail emporium of Victoria, the capital of British Columbia, is the Hudson Bay store; and in Calgary, the metropolis of Alberta and the Canadian plains, the principal shopping-place in a territory beside which Texas dwindles to the proportions of a park is the Hudson Bay store.

These and many other shops indicate a new development of the business of the last of England's great chartered monopolies; but instead of marking the manner in which civilization has forced it to abandon its original function, this merely demonstrates that the proprietors have taken advantage of new conditions while still pursuing their original trade. It is true that the huge corporation is becoming a great retail shop-keeping company. It is also true that by the surrender of its monopolistic privileges it got a consolation prize of money and of twenty millions of dollars' worth of land, so that its chief business may yet become that of developing and selling real estate. But to-day it is still, as it was two centuries ago, the greatest of fur-trading corporations, and fur-trading is to-day a principal source of its profits.

Reminders of their old associations as forts still confront the visitor to the modern city shops of the company. The great shop in Victoria, for instance, which, as a fort, was the hub around which grew the wheel that is now the capital of the province, has its fur trade conducted in a sort of barn-like annex of the bazaar; but there it is, nevertheless, and busy among the great heaps of furs are men who can remember when the Hydahs and the T'linkets and the other neighboring tribes came down in their war canoes to trade their winter's catch of skins for guns and beads, vermilion, blankets, and the rest. Now this is the mere catch-all for the furs got at posts farther up the coast and in the interior. But upstairs, above the store, where the fashionable ladies are looking over laces and purchasing perfumes, you will see a collection of queer old guns of a pattern familiar to Daniel Boone. They are relics of the fur company's stock of those famous "trade-guns" which disappeared long before they had cleared the plains of buffalo, and which the Indians used to deck with brass nails and bright paint, and value as no man to-day values a watch. But close to the trade-guns of romantic memory is something yet more highly suggestive of the company's former position. This is a heap of unclaimed trunks, "left," the employés will tell you, "by travellers, hunters, and explorers who never came back to inquire for them."

It was not long ago that conditions existed such as in that region rendered the disappearance of a traveller more than a possibility. The wretched, squat, bow-legged, dirty laborers of that coast, who now dress as we do, and earn good wages in the salmon-fishing and canning industries, were not long ago very numerous, and still more villanous. They were not to be compared with the plains Indians as warriors or as men, but they were more treacherous, and wanting in high qualities. In the interior to-day are some Indians such as they were who are accused of cannibalism, and who have necessitated warlike defences at distant trading-posts. Travellers who escaped Indian treachery risked starvation, and stood their chances of losing their reckoning, of freezing to death, of encounters with grizzlies, of snow-slides, of canoe accidents in rapids, and of all the other casualties of life in a territory which to-day is not half explored. Those are not the trunks of Hudson Bay men, for such would have been sent home to English and Scottish mourners; they are the luggage of chance men who happened along, and outfitted at the old post before going farther. But the company's men were there before them, had penetrated the

region farther and earlier, and there they are to-day, carrying on the fur trade under conditions strongly resembling those their predecessors once encountered at posts that are now towns in farming regions, and where now the locomotive and the steamer are familiar vehicles. Moreover, the status of the company in British Columbia is its status all the way across the North from the Pacific to the Atlantic.

To me the most interesting and picturesque life to be found in North America, at least north of Mexico, is that which is occasioned by this principal phase of the company's operations. In and around the fur trade is found the most notable relic of the white man's earliest life on this continent. Our wild life in this country is, happily, gone. The frontiersman is more difficult to find than the frontier, the cowboy has become a laborer almost like any other, our Indians are as the animals in our parks, and there is little of our country that is not threaded by railroads or wagon-ways. But in new or western Canada this is not so. A vast extent of it north of the Canadian Pacific Railway, which hugs our border, has been explored only as to its waterways, its valleys, or its open plains, and where it has been traversed much of it remains as Nature and her near of kin, the red men, had it of old. On the streams canoes are the vehicles of travel and of commerce; in the forests "trails" lead from trading-post to trading-post, the people are Indians, half-breeds, and Esquimaux, who live by hunting and fishing as their forebears did; the Hudson Bay posts are the seats of white population; the post factors are the magistrates.

All this is changing with a rapidity which history will liken to the sliding of scenes before the lens of a magic-lantern. Miners are crushing the foot-hills on either side of the Rocky Mountains, farmers and cattle-men have advanced far northward on the prairie and on the plains in narrow lines, and railroads are pushing hither and thither. Soon the limits of the inhospitable zone this side of the Arctic Sea, and of the marshy, weakly-wooded country on either side of Hudson Bay will circumscribe the fur-trader's field, except in so far as there may remain equally permanent hunting-grounds in Labrador and in the mountains of British Columbia. Therefore now, when the Hudson Bay Company is laying the foundations of widely different interests, is the time for halting the old original view that stood in the stereopticon for centuries, that we may see what it revealed, and will still show far longer than it takes for us to view it.

The Hudson Bay Company's agents were not the first hunters and fur-traders in British America, ancient as was their foundation. The French, from the Canadas, preceded them no one knows how many years, though it is said that it was as early as 1627 that Louis XIII. chartered a company of the same sort and for the same aims as the English company. Whatever came of that corporation I do not know, but by the time the Englishmen established themselves on Hudson Bay, individual Frenchmen and half-breeds had penetrated the country still farther west. They were of hardy, adventurous stock, and they loved the free roving life of the trapper and hunter. Fitted out by the merchants of Canada, they would pursue the waterways which there cut up the wilderness in every direction, their canoes laden with goods to tempt the savages, and their guns or traps forming part of their burden. They would be gone the greater part of a year, and always returned with a store of furs to be converted into money, which was, in turn, dissipated in the cities with devil-may-care jollity. These were the *coureurs du bois*, and theirs was the stock from which came the *voyageurs* of the next era, and the half-breeds, who joined the service of the rival fur companies, and who, by-the-way, reddened the history of the North-west territories with the little bloodshed that mars it.

Charles II. of England was made to believe that wonders in the way of discovery and trade would result from a grant of the Hudson Bay territory to certain friends and petitioners. An experimental voyage was made with good results in 1668, and in 1670 the King granted the charter to what he styled "the Governor and Company of Adventurers of England trading into Hudson's Bay, one body corporate and politique, in deed and in name, really and fully forever, for Us, Our heirs, and Successors." It was indeed a royal and a wholesale charter, for the King declared, "We have given, granted, and confirmed unto said Governor and Company sole trade and commerce of those Seas, Streights, Bays, Rivers, Lakes, Creeks, and Sounds, in whatsoever latitude they shall be, that lie within the entrance of the Streights commonly called Hudson's, together with all the Lands, Countries, and Territories upon the coasts and confines of the Seas, etc., ... not already actually possessed by or granted to any of our subjects, or possessed by the subjects of any other Christian Prince or State, with the fishing of all sorts of Fish, Whales, Sturgeons, and all other Royal Fishes, ... together with the Royalty of the Sea upon the Coasts within the limits aforesaid, and all Mines Royal, as well discovered as not discovered, of Gold, Silver, Gems, and Precious Stones, ... and that the said lands be henceforth reckoned and reputed as one of Our Plantations or Colonies in America called Rupert's Land." For this gift of an empire the corporation was to pay yearly to the king, his heirs and successors, two elks and two black beavers whenever and as often as he, his heirs, or his successors "shall happen to enter into the said countries." The company was empowered to man ships of war, to create an armed force for security and defence, to make peace or war with any people that were not Christians, and to seize any British or other subject who traded in their territory. The King named his cousin, Prince Rupert, Duke of Cumberland, to be first governor, and it was in his honor that the new territory got its name of Rupert's Land.

In the company were the Duke of Albemarle, Earl Craven, Lords Arlington and Ashley, and several knights and baronets, Sir Philip Carteret among them. There were also five esquires, or gentlemen, and John Portman, "citizen and goldsmith." They adopted the witty sentence, "*Pro pelle cutem*" (A skin for a skin), as their motto, and established as their coat of arms a fox sejant as the crest, and a

THE BEAR TRAP

shield showing four beavers in the quarters, and the cross of St. George, the whole upheld by two stags.

The "adventurers" quickly established forts on the shores of Hudson Bay, and began trading with the Indians, with such success that it was rumored they made from twenty-five to fifty per cent. profit every year. But they exhibited all of that timidity which capital is ever said to possess. They were nothing like as enterprising as the French *coureurs du bois*. In a hundred years they were no deeper in the country than at first, excepting as they extended their little system of forts or "factories" up and down and on either side of Hudson and James bays. In view of their profits, perhaps this lack of enterprise is not to be wondered at. On the other hand, their charter was given as a reward for the efforts they had made, and were to make, to find "the Northwest passage to the Southern seas." In this quest they made less of a trial than in the getting of furs; how much less we shall see. But the company had no lack of brave and hardy followers. At first many of the men at the factories were from the Orkney Islands, and those islands remained until recent times the recruiting-source for this service. This was because the Orkney men were inured to a rigorous climate, and to a diet largely composed of fish. They were subject to less of a change in the company's service than must have been endured by men from almost any part of England.

I am going, later, to ask the reader to visit Rupert's Land when the company had shaken off its timidity, overcome its obstacles, and dotted all British America with its posts and forts. Then we shall see the interiors of the forts, view the strange yet not always hard or uncouth life of the company's factors and clerks; and glance along the trails and watercourses, mainly unchanged to-day, to note the work and surroundings of the Indians, the *voyageurs*, and the rest who inhabit that region. But, fortunately, I can first show, at least roughly, much that is interesting about the company's growth and methods a century and a half ago. The information is gotten from some

English Parliamentary papers forming a report of a committee of the House of Commons in 1749.

Arthur Dobbs and others petitioned Parliament to give them either the rights of the Hudson Bay Company or a similar charter. It seems that England had offered £20,000 reward to whosoever should find the bothersome passage to the Southern seas *via* this northern route, and that these petitioners had sent out two ships for that purpose. They said that when others had done no more than this in Charles II.'s time, that monarch had given them " the greatest privileges as lords proprietors " of the Hudson Bay territory, and that those recipients of royal favor were bounden to attempt the discovery of the desired passage. Instead of this, they not only failed to search effectually or in earnest for the passage, but they had rather endeavored to conceal the same, and to obstruct the discovery thereof by others. They had not possessed or occupied any of the lands granted to them, or extended their trade, or made any plantations or settlements, or permitted other British subjects to plant, settle, or trade there. They had established only four factories and one small trading-house; yet they had connived at or allowed the French to encroach, settle, and trade within their limits, to the great detriment and loss of Great Britain. The petitioners argued that the Hudson Bay charter was monopolistic, and therefore void, and at any rate it had been forfeited " by non-user or abuser."

In the course of the hearing upon both sides, the " voyages upon discovery," according to the company's own showing, were not undertaken until the corporation had been in existence nearly fifty years, and then the search had only been prosecuted during eighteen years, and with only ten expeditions. Two ships sent out from England never reached the bay, but those which succeeded, and were then ready for adventurous cruising, made exploratory voyages that lasted only between one month and ten weeks, so that, as we are accustomed to judge such expeditions, they seem farcical and mere pretences. Yet their largest ship was only of 190 tons burden, and the others were a third smaller — vessels like our small coasting schooners. The most particular instructions to the captains were to trade with all natives, and persuade them to kill whales, sea-horses, and seals; and, subordinately and incidentally, "by God's permission," to find out the Strait of Annian, a fanciful sheet of water, with tales of which that irresponsible Greek sea-tramp, Juan de Fuca, had disturbed all Christendom, saying that it led between a great island in the Pacific (Vancouver) and the mainland into the inland lakes. To the factors at their forts the company sent such lukewarm messages as, "and if you can by any means find out any discovery or matter to the northward or elsewhere in the company's interest or advantage, do not fail to let us know every year."

The attitude of the company towards discovery suggests a Dogberry at its head, bidding his servants to "comprehend" the North-west passage, but should they fail, to thank God they were rid of a villain. In truth, they were traders pure and simple, and were making great profits with little trouble and expense.

They brought from England about £4000 worth of powder, shot, guns, fire-steels, flints, gun-worms, powder-horns, pistols, hatchets, sword blades, awl blades, ice-chisels, files, kettles, fish-hooks, net-lines, burning-glasses, looking-glasses, tobacco, brandy, goggles, gloves, hats, lace, needles, thread, thimbles, breeches, vermilion, worsted sashes, blankets, flannels, red feathers, buttons, beads, and "shirts, shoes, and stockens." They spent, in keeping up their posts

HUSKIE DOGS FIGHTING

and ships, about £15,000, and in return they brought to England castorum, whale-fins, whale-oil, deer-horns, goose-quills, bed-feathers, and skins—in all of a value of about £26,000 per annum. I have taken the average for several years in that period of the company's history, and it is in our money as if they spent $90,000 and got back $130,000, and this is their own showing under such circumstances as to make it the course of wisdom not to boast of their profits. They had three times trebled their stock and otherwise increased it, so that having been 10,500 shares at the outset, it was now 103,950 shares.

And now that we have seen how natural it was that they should not then bother with exploration and discovery, in view of the remuneration that came for simply sitting in their forts and buying furs, let me pause to repeat what one of their wisest men said casually, between the whiffs of a meditative cigar, last summer: "The search for the north pole must soon be taken up in earnest," said he. "Man has paused in the undertaking because other fields where his needs were more pressing, and where effort was more certain to be rewarded with success, had been neglected. This is no longer the fact, and geographers and other students of the subject all agree that the north pole must next be sought and found. Speaking only on my own account and from my knowledge, I assert that whenever any government is in earnest in this desire, it will employ the men of this fur service, and they will find the pole. The company has posts far within the arctic circle, and they are manned by men peculiarly and exactly fitted for the adventure. They are hardy, acutely intelligent, self-reliant, accustomed to the climate, and all that it engenders and demands. They are on the spot ready to start at the earliest moment in the season, and they have with them all that they will need on the expedition. They would do nothing hurriedly or rashly; they would know what they were about as no other white men would — and they would get there."

I mention this not merely for the novelty of the suggestion and the interest it may excite, but because it contributes to the reader's understanding of the scope and character of the work of the company. It is not merely Western and among Indians, it is hyperborean and among Esquimaux. But would it not be passing strange if, beyond all that England has gained from the careless gift of an empire to a few favorites by Charles II., she should yet possess the honor and glory of a grand discovery due to the natural results of that action?

To return to the Parliamentary inquiry into the company's affairs 140 years ago. If it served no other purpose, it drew for us of this day an outline picture of the first forts and their inmates and customs. Being printed in the form our language took in that day, when a gun was a "musquet" and a stockade was a "palisadoe," we fancy we can see the bumptious governors — as they then called the factors or agents — swelling about in knee-breeches and cocked hats and colored waistcoats, and relying, through their fear of the savages, upon the little putty-pipe cannon that they speak of as "swivels."

These were ostentatiously planted before their quarters, and in front of these again were massive double doors, such as we still make of steel for our bank safes, but, when made of wood, use only for our refrigerators. The views we get of the company's "servants"—which is to say, mechanics and laborers—are all of trembling varlets, and the testimony is full of hints of petty sharp practice towards the red man, suggestive of the artful ways of our own Hollanders, who bought beaver-skins by the weight of their feet, and then pressed down upon the scales with all their might.

The witnesses had mainly been at one time in the employ of the company, and they made the point against it that it imported all its bread (*i. e.*, grain) from England, and neither encouraged planting nor cultivated the soil for itself. But there were several who said that even in August they found the soil still frozen at a depth of two and a half or three feet. Not a man in the service was allowed to trade with the natives outside the forts, or even to speak with them. One fellow was put in irons for going into an Indian's tent; and there was a witness who had "heard a Governor say he would whip a Man without Tryal; and that the severest Punishment is a Dozen of Lashes." Of course there was no instructing the savages in either English or the Christian religion; and we read that, though there were twenty-eight Europeans in one factory, "witness never heard Sermon or Prayers there, nor ever heard of any such Thing either before his Time or since." Hunters who offered their services got one-half what they

shot or trapped, and the captains of vessels kept in the bay were allowed "25 *l. per cent.*" for all the whalebone they got.

One witness said: "The method of trade is by a standard set by the Governors. They never lower it, but often double it, so that where the Standard directs 1 Skin to be taken they generally take Two." Another said he "had been ordered to shorten the measure for Powder, which ought to be a Pound, and that within these 10 Years had been reduced an Ounce or Two." "The Indians made a Noise sometimes, and the Company gave them their Furs again." A book-keeper lately in the service said that the company's measures for powder were short, and yet even such measures were not filled above half full. Profits thus made were distinguished as "the overplus trade," and signified what skins were got more than were paid for, but he could not say whether such gains went to the company or to the governor. (As a matter of fact, the factors or governors shared in the company's profits, and were interested in swelling them in every way they could.)

There was much news of how the French traders got the small furs of martens, foxes, and cats, by intercepting the Indians, and leaving them to carry only the coarse furs to the company's forts. A witness "had seen the Indians come down in fine *French* cloaths, with as much Lace as he ever saw upon any Cloaths whatsoever. He believed if the Company would give as much for the Furs as the *French*, the *Indians* would bring them down;" but the French asked only thirty marten-skins for a gun, whereas the company's standard was from thirty-six to forty such skins. Then, again, the company's plan (unchanged to-day) was to take the Indian's furs, and then, being possessed of them, to begin the barter.

This shouldering the common grief upon the French was not merely the result of the chronic English antipathy to their ancient and their lively foes. The French were swarming all around the outer limits of the company's field, taking first choice of the furs, and even beginning to set up posts of their own. Canada was French soil, and peopled by as hardy and adventurous a class as inhabited any part of America. The *coureurs du bois* and the *bois-brûlés* (half-breeds), whose success afterwards led to the formation of rival companies, had begun a mosquito warfare, by canoeing the waters that led to Hudson Bay, and had penetrated 1000 miles farther west than the English. One Thomas Barnett, a smith, said that the French intercepted the Indians, forcing them to trade, " when they take what they please, giving them Toys in Exchange; and fright them into Compliance by Tricks of Sleight of Hand; from whence the *Indians* conclude them to be Conjurers; and if the *French* did not compel the *Indians* to trade, they would certainly bring all the Goods to the *English*."

This must have seemed to the direct, practical English trading mind a wretched business, and worthy only of Johnny Crapeau, to worst the noble Briton by monkeyish acts of conjuring. It stirred the soul of one witness, who said that the way to meet it was " by sending some *English* with a little Brandy."

A gallon to certain chiefs and a gallon and a half to others would certainly induce the natives to come down and trade, he thought.

But while the testimony of the English was valuable as far as it went, which was mainly concerning trade, it was as nothing regarding the life of the natives compared with that of one Joseph La France, of Missili-Mackinack (Mackinaw), a traveller, hunter, and trader. He had been sent as a child to Quebec to learn French, and in later years had been from Lake Nipissing to Lake Champlain and the Great Lakes, the Mississippi, the Missouri, the Ouinipigue (Winnipeg) or Red River, and to Hudson Bay. He told his tales to Arthur Dobbs, who made a book of them, and part of that became an appendix to the committee's report. La France said:

"That the high price on *European* Goods discourages the Natives so much, that if it were not that they are under a Necessity of having Guns, Powder, Shot, Hatchets, and other Iron Tools for their Hunting, and Tobacco, Brandy, and some Paint for Luxury, they would not go down to the Factory with what they now carry. They leave great numbers of Furs and Skins behind them. A good Hunter among the *Indians* can kill 600 Beavers in a season, and carry down but 100" (because their canoes were small); "the rest he uses at home, or hangs them upon Branches of Trees upon the Death of their Children, as an Offering to them; or use them for Bedding and Coverings; they sometimes burn off the Fur, and roast the Beavers, like Pigs, upon any Entertainments; and they often let them rot, having no further Use of them. The Beavers, he says, are of Three Colours—the Brown-reddish Colour, the Black, and the White. The Black is most valued by the Company, and in *England*; the White, though most valued in *Canada*, is blown upon by the Company's Factors at the Bay, they not allowing so much for these as for the others; and therefore the *Indians* use them at home, or burn off the Hair, when they roast the Beavers, like Pigs, at an Entertainment when they feast together. The Beavers are delicious Food, but the

Tongue and Tail the most delicious Parts of the whole. They multiply very fast, and if they can empty a Pond, and take the whole Lodge, they generally leave a Pair to breed, so that they are fully stocked again in Two or Three Years. The *American* Oxen, or Beeves, he says, have a large Bunch upon their backs, which is by far the most delicious Part of them for Food, it being all as sweet as Marrow, juicy and rich, and weighs several Pounds.

"The Natives are so discouraged in their Trade with the Company that no Peltry is worth the Carriage; and the finest Furs are sold for very little. They gave but a Pound of Gunpowder for 4 Beavers, a Fathom of Tobacco for 7 Beavers, a Pound of Shot for 1, an Ell of coarse Cloth for 15, a Blanket for 12, Two Fish-hooks or Three Flints for 1; a Gun for 25, a Pistol for 10, a common Hat with white Lace, 7; an Ax, 4; a Billhook, 1; a Gallon of Brandy, 4; a chequer'd Shirt, 7; all of which are sold at a monstrous Profit, even to 2000 *per Cent*. Notwithstanding this discouragement, he computed that there were brought to the Factory in 1742, in all, 50,000 Beavers and above 9000 Martens.

"The smaller Game, got by Traps or Snares, are generally the Employment of the Women and Children; such as the Martens, Squirrels, Cats, Ermines, &c. The Elks, Stags, Rein-Deer, Bears, Tygers, wild Beeves, Wolves, Foxes, Beavers, Otters, Corcajeu, &c., are the employment of the Men. The *Indians*, when they kill any Game for Food, leave it where they kill it, and send their wives next Day to carry it home. They go home in a direct Line, never missing their way, by observations they make of the Course they take upon their going out. The Trees all bend towards the South, and the Branches on that Side are larger and stronger than on the North Side; as also the Moss upon the Trees. To let their Wives know how to come at the killed Game, they from Place to Place break off Branches and lay them in the Road, pointing them the Way they should go, and sometimes Moss; so that they never miss finding it.

"In Winter, when they go abroad, which they must do in all Weathers, before they dress, they rub themselves all over with Bears Greaze or Oil of Beavers, which does not freeze; and also rub all the Fur of their Beaver Coats, and then put them on; they have also a kind of Boots or Stockings of Beaver's Skin, well oiled, with the Fur inwards; and above them they have an oiled Skin laced about their Feet, which keeps out the Cold, and also Water; and by this means they never freeze, nor suffer anything by Cold. In Summer, also, when they go naked, they rub themselves with these Oils or Grease, and expose themselves to the Sun without being scorched, their Skins always being kept soft and supple by it; nor do any Flies, Bugs, or Musketoes, or any noxious Insect, ever molest them. When they want to get rid of it, they go into the Water, and rub themselves all over with Mud or Clay, and let it dry upon them, and then rub it off; but whenever they are free from the Oil, the Flies and Musketoes immediately attack them, and oblige them again to anoint themselves. They are much afraid of the wild Humble Bee, they going naked in Summer, that they avoid them as much as they can. They use no Milk from the time they are weaned, and they all hate to taste Cheese, having taken up an Opinion that it is made of Dead Men's Fat. They love Prunes and Raisins, and will give a Beaver-skin for Twelve of them, to carry to their Children; and also for a Trump or Jew's Harp. The Women have all fine Voices, but have never heard any Musical Instrument. They are very fond of all Kinds of Pictures or Prints, giving a Beaver for the least Print; and all Toys are like Jewels to them."

He reported that "the *Indians* west of Hudson's Bay live an erratic Life, and can have no Benefit by tame Fowl or Cattle. They seldom stay above a Fortnight in a Place, unless they find Plenty of Game. After having built their Hut, they disperse to get Game for their Food, and meet again at Night after having killed enough to maintain them for that Day. When they find Scarcity of Game, they remove a League or Two farther; and thus they traverse through woody Countries and Bogs, scarce missing One Day, Winter or Summer, fair or foul, in the greatest Storms of Snow."

It has been often said that the great Peace River, which rises in British Columbia and flows through a pass in the Rocky Mountains into the northern plains, was named "the Unchaga," or Peace, "because" (to quote Captain W. F. Butler) "of the stubborn resistance offered by the all-conquering Crees, which induced that warlike tribe to make peace on the banks of the river, and leave at rest the beaver-hunters"—that is, the Beaver tribe—upon the river's banks. There is a sentence in La France's story that intimates a more probable and lasting reason for the name. He says that some Indians in the southern centre of Canada sent frequently to the Indians along some river near the mountains "with presents, to confirm the peace with them." The story is shadowy, of course, and yet La France, in the same narrative, gave other information which proved to be correct, and none which proved ridiculous. We know that there were "all-conquering" Crees, but there were also inferior ones called the Swampies, and there were others of only intermediate valor. As for the Beavers, Captain Butler himself offers other proof of their mettle besides their "stubborn resistance." He says that on one occasion a young Beaver chief shot the dog of another brave in the Beaver camp. A hundred bows were instantly drawn, and ere night eighty of the best men of the tribe lay dead. There was a parley, and it was resolved that the chief who slew the dog should leave the tribe, and take his friends with him. A century later a Beaver Indian, travelling with a white man, heard his own tongue spoken by men among the Blackfeet near our border. They were the Sarcis, descendants of the exiled band of Beavers. They had become the most reckless and valorous members of the warlike Blackfeet confederacy.

La France said that the nations who "go up the river" with presents, to confirm the peace with certain Indians, were three months in going, and that

COUREUR DU BOIS

the Indians in question live beyond a range of mountains beyond the Assiniboins (a plains tribe). Then he goes on to say that still farther beyond those Indians "are nations who have not the use of firearms, by which many of them are made slaves and sold"—to the Assiniboins and others. These are plainly the Pacific coast Indians. And even so long ago as that (about 1740), half a century before Mackenzie and Vancouver met on the Pacific coast, La France had told the story of an Indian who had gone at the head of a band of thirty braves and their families to make war on the Flatheads "on the Western Ocean of America." They were from autumn until the next April in making the journey, and they "saw many Black Fish spouting up in the sea." It was a case of what the Irish call "spoiling for a fight," for they had to journey 1500 miles to meet "enemies" whom they never had seen, and who were peaceful, and inhabited more or less permanent villages. The plainsmen got more than they sought. They attacked a village, were outnumbered, and lost half their force, besides having several of their men wounded. On the way back all except the man who told the story died of fatigue and famine.

The journeys which Indians made in their wildest period were tremendous. Far up in the wilderness of British America there are legends of visits by the Iroquois. The Blackfeet believe that their progenitors roamed as far south as Mexico for horses, and the Crees of the plains evinced a correct knowledge of the country that lay beyond the Rocky Mountains in their conversations with the first whites who traded with them. Yet those white men, the founders of an organized fur trade, clung to the scene of their first operations for more than one hundred years, while the bravest of their more enterprising rivals in the Northwest Company only reached the Pacific, with the aid of eight Iroquois braves, 120 years after the English king chartered the senior company! The French were the true Yankees of that country.

They and their half-breeds were always in the van as explorers and traders, and as early as 1731 M. Varennes de la Verandrye, licensed by the Canadian Government as a trader, penetrated the West as far as the Rockies, leading Sir Alexander Mackenzie to that extent by more than sixty years.

But to return to the first serious trouble the Hudson Bay Company met. The investigation of its affairs by Parliament produced nothing more than the picture I have presented. The committee reported that if the original charter bred a monopoly, it would not help matters to give the same privileges to others. As the questioned legality of the charter was not competently adjudicated upon, they would not allow another company to invade the premises of the older one.

At this time the great company still hugged the shores of the bay, fearing the Indians, the half-breeds, and the French. Their posts were only six in all, and were mainly fortified with palisaded enclosures, with howitzers and swivels, and with men trained to the use of guns. Moose Fort and the East Main factory were on either side of James Bay, Forts Albany, York, and Prince of Wales followed up the west coast, and Henley was the southernmost and most inland of all, being on Moose River, a tributary of James Bay. The French at first traded beyond the field of Hudson Bay operations, and their castles were their canoes. But when their great profits and familiarity with the trade tempted the thrifty French capitalists and enterprising Scotch merchants of Montreal into the formation of the rival Northwest

Trading Company in 1783, fixed trading-posts began to be established all over the Prince Rupert's Land, and even beyond the Rocky Mountains in British Columbia. By 1818 there were about forty Northwest posts as against about two dozen Hudson Bay factories. The new company not only disputed but ignored the chartered rights of the old company, holding that the charter had not been sanctioned by Parliament, and was in every way unconstitutional as creative of a monopoly. Their French partners and *engagés* shared this feeling, especially as the French crown had been first in the field with a royal charter. Growing bolder and bolder, the Northwest Company resolved to drive the Hudson Bay Company to a legal test of their rights, and so in 1803-4 they established a Northwest fort under the eyes of the old company on the shores of Hudson Bay, and fitted out ships to trade with the natives in the strait. But the Englishmen did not accept the challenge; for the truth was they had their own doubts of the strength of their charter.

They pursued a different and for them an equally bold course. That hard-headed old nobleman the fifth Earl of Selkirk came uppermost in the company as the engineer of a plan of colonization. There was plenty of land, and some wholesale evictions of Highlanders in Sutherlandshire, Scotland, had rendered a great force of hardy men homeless. Selkirk saw in this situation a chance to play a long but certainly triumphant game with his rivals. His plan was to plant a colony which should produce grain and horses and men for the old company, saving the im-

portation of all three, and building up not only a nursery for men to match the *coureurs du bois*, but a stronghold and a seat of a future government in the Hudson Bay interest. Thus was ushered in a new and important era in Canadian history. It was the opening of that part of Canada; by a loop-hole rather than a door, to be sure.

Lord Selkirk's was a practical soul. On one occasion in animadverting against the Northwest Company he spoke of them contemptuously as fur-traders, yet he was the chief of all fur-traders, and had been known to barter with an Indian himself at one of the forts for a fur. He held up the opposition to the scorn of the world as profiting upon the weakness of the Indians by giving them alcohol, yet he ordered distilleries set up in his colony afterwards, saying, "We grant the trade is iniquitous, but if we don't carry it on others will; so we may as well put the guineas in our own pockets." But he was the man of the moment, if not for it. His scheme of colonization was born of desperation on one side and distress on the other. It was pursued amid terrible hardship, and against incessant violence. It was consummated through bloodshed. The story is as interesting as it is important. The facts are obtained mainly from "Papers relating to the Red River Settlement, ordered to be printed by the House of Commons, July 12, 1819." Lord Selkirk owned 40,000 of the £105,000 (or shares) of the Hudson Bay Company; therefore, since 25,000 were held by women and children, he held half of all that carried votes. He got from the company a grant of a large tract around what is now Winnipeg, to form an agricultural settlement for supplying the company's posts with provisions. We have seen how little disposed its officers were to open the land to settlers, or to test its agricultural capacities. No one, therefore, will wonder that when this grant was made several members of the governing committee resigned. But a queer development of the moment was a strong opposition from holders of Hudson Bay stock who were also owners in that company's great rival, the Northwest Company. Since the enemy persisted in prospering at the expense of the old company, the moneyed men of the senior corporation had taken stock of their rivals. These doubly interested persons were also in London, so that the Northwest Company was no longer purely Canadian. The opponents within the Hudson Bay Company declared civilization to be at all times unfavorable to the fur trade, and the Northwest people argued that the colony would form a nursery for servants of the Bay Company, enabling them to oppose the Northwest Company more effectually, as well as affording such facilities for new-comers as must destroy their own monopoly. The Northwest Company denied the legality of the charter rights of the Hudson Bay Company because Parliament had not confirmed Charles II.'s charter.

The colonists came, and were met by Miles McDonnell, an ex-captain of Canadian volunteers, as Lord Selkirk's agent. The immigrants landed on the shore of Hudson Bay, and passed a forlorn winter. They met some of the Northwest Company's people

under Alexander McDonnell, a cousin and brother-in-law to Miles McDonnell. Although Captain Miles read the grant to Selkirk in token of his sole right to the land, the settlers were hospitably received and well treated by the Northwest people. The settlers reached the place of colonization in August, 1812. This place is what was known as Fort Garry until Winnipeg was built. It was at first called "the Forks of the Red River," because the Assiniboin there joined the Red. Lord Selkirk outlined his policy at the time in a letter in which he bade Miles McDonnell give the Northwest people solemn warning that the lands were Hudson Bay property, and they must remove from them; that they must not fish, and that if they did their nets were to be seized, their buildings were to be destroyed, and they were to be treated "as you would poachers in England."

The trouble began at once. Miles accused Alexander of trying to inveigle colonists away from him. He trained his men in the use of guns, and uniformed a number of them. He forbade the exportation of any supplies from the country, and when some Northwest men came to get baffalo meat they had hung on racks in the open air, according to the custom of the country, he sent armed men to send the others away. He intercepted a band of Northwest canoe-men, stationing men with guns and with two field-pieces on the river; and he sent to a Northwest post lower down the river demanding the provisions stored there, which, when they were refused, were taken by force, the door being smashed in. For this a Hudson Bay clerk was arrested, and Captain Miles's men went to the rescue. Two armed forces met, but happily slaughter was averted. Miles McDonnell justified his course on the ground that the colonists were distressed by need of food. It transpired at the time that one of his men while making cartridges for a cannon remarked that he was making them "for those —— Northwest rascals. They have run too long, and shall run no longer." After this Captain Miles ordered the stoppage of all buffalo-hunting on horseback, as the practice kept the buffalo at a distance, and drove them into the Sioux country, where the local Indians dared not go.

But though Captain McDonnell was aggressive and vexatious, the Northwest Company's people, who had begun the mischief, even in London, were not now passive. They relied on setting the half-breeds and Indians against the colonists. They urged that the colonists had stolen Indian real estate in settling on the land, and that in time every Indian would starve as a consequence. At the forty-fifth annual meeting of the Northwest Company's officers, August, 1814, Alexander McDonnell said, "Nothing but the complete downfall of the colony will satisfy some, by fair or foul means—a most desirable object, if it can be accomplished; so here is at it with all my heart and energy." In October, 1814, Captain McDonnell ordered the Northwest Company to remove from the territory within six months.

The Indians, first and last, were the friends of the colonists. They were befriended by the whites, and in turn they gave them succor when famine fell upon them. Many of Captain Miles McDonnell's orders were in their interest, and they knew it. Katawabetay, a chief, was tempted with a big prize to destroy the settlement. He refused. On the opening of navigation in 1815 chiefs were bidden from the country around to visit the Northwest factors, and were by them asked to destroy the colony. Not only

THE INDIAN HUNTER OF 1750

did they decline, but they hastened to Captain Miles McDonnell to acquaint him with the plot. Duncan Cameron now appears foremost among the Northwest Company's agents, being in charge of that company's post on the Red River, in the Selkirk grant.

He told the chiefs that if they took the part of the colonists "their camp-fires should be totally extinguished." When Cameron caught one of his own servants doing a trifling service for Captain Miles McDonnell, he sent him upon a journey for which every *engagé* of the Northwest Company bound himself liable in joining the company; that was to make the trip to Montreal, a voyage held *in terrorem* over every servant of the corporation. More than that, he confiscated four horses and a wagon belonging to this man, and charged him on the company's books with the sum of 800 livres for an Indian squaw, whom the man had been told he was to have as his slave for a present.

But though the Indians held aloof from the great and cruel conspiracy, the half-breeds readily joined in it. They treated Captain McDonnell's orders with contempt, and arrested one of the Hudson Bay men as a spy upon their hunting with horses. There lived along the Red River, near the colony, about thirty Canadians and seventy half-breeds, born of Indian squaws and the servants or officers of the Northwest Company. One-quarter of the number of "breeds" could read and write, and were fit to serve as clerks; the rest were literally half savage, and were employed as hunters, canoe-men, "packers" (freighters), and guides. They were naturally inclined to side with the Northwest Company, and in time that corporation sowed dissension among the colonists themselves, picturing to them exaggerated danger from the Indians, and offering them free passage to Canada. They paid at least one of the lead-

INDIAN HUNTER HANGING DEER OUT OF THE REACH OF WOLVES

ing colonists £100 for furthering discontent in the settlement, and four deserters from the colony stole all the Hudson Bay field-pieces, iron swivels, and the howitzer. There was constant irritation and friction between the factions. In an affray far up at Isle-à-la-Crosse a man was killed on either side. Half-breeds came past the colony singing war-songs, and notices were posted around Fort Garry reading, "Peace with all the world except in Red River." The Northwest people demanded the surrender of Captain McDonnell that he might be tried on their charges, and on June 11, 1815, a band of men fired on the colonial buildings. The captain afterwards surrendered himself, and the remnant of the colony, thirteen families, went to the head of Lake Winnipeg. The half-breeds burned the buildings, and divided the horses and effects.

But in the autumn all came back with Colin Robertson, of the Bay Company, and twenty clerks and servants. These were joined by Governor Robert Semple, who brought 160 settlers from Scotland. Semple was a man of consequence at home, a great traveller, and the author of a book on travels in Spain.* But he came in no conciliatory mood, and the foment was kept up. The Northwest Company

---

* I am indebted to Mr. Matthew Semple, of Philadelphia, a grand-nephew of the murdered Governor, for further facts about that hero. He led a life of travel and adventure, spiced with almost romantic happenings. He wrote ten books: records of travel and one novel. His parents were passengers on an English vessel which was captured by the Americans in 1776, and brought to Boston, Mass., where he was born on February 26, 1777. He was therefore only 39 years of age when he was slain. His portrait, now in Philadelphia, shows him to have been a man of striking and handsome appearance.

tried to starve the colonists, and Governor Semple destroyed the enemy's fort below Fort Garry. Then came the end—a decisive battle and massacre.

Sixty-five men on horses, and with some carts, were sent by Alexander McDonnell, of the Northwest Company, up the river towards the colony. They were led by Cuthbert Grant, and included six Canadians, four Indians, and fifty-four half-breeds. It was afterwards said they went on innocent business, but every man was armed, and the "breeds" were naked, and painted all over to look like Indians. They got their paint of the Northwest officers. Moreover, there had been rumors that the colonists were to be driven away, and that "the land was to be drenched with blood." It was on June 19, 1816, that runners notified the colony that the others were coming. Semple was at Fort Douglas, near Fort Garry. When apprised of the close approach of his assailants, the Governor seems not to have appreciated his danger, for he said, "We must go and meet those people; let twenty men follow me." He put on his cocked hat and sash, his pistols, and shouldered his double-barrelled fowling-piece. The others carried a wretched lot of guns—some with the locks gone, and many that were useless. It was marshy ground, and they straggled on in loose order. They met an old soldier who had served in the army at home, and who said the enemy was very numerous, and that the Governor had better bring along his two field-pieces.

"No, no," said the Governor; "there is no occasion. I am only going to speak to them."

Nevertheless, after a moment's reflection, he did send back for one of the great guns, saying it was well to have it in case of need. They halted a short time for the cannon, and then perceived the Northwest party pressing towards them on their horses. By a common impulse the Governor and his followers began a retreat, walking backwards, and at the same time spreading out a single line to present a longer front. The enemy continued to advance at a hand-gallop. From out among them rode a Canadian named Boucher, the rest forming a half-moon behind him. Waving his hand in an insolent way to the Governor, Boucher called out, "What do you want?"

MAKING THE SNOW-SHOE.

"What do *you* want?" said Governor Semple.

"We want our fort," said Boucher, meaning the fort Semple had destroyed.

"Go to your fort," said the Governor.

"Why did you destroy our fort, you rascal?" Boucher demanded.

"Scoundrel, do you tell me so?" the Governor replied, and ordered the man's arrest.

Some say he caught at Boucher's gun. But Boucher slipped off his horse, and on the instant a gun was fired, and a Hudson Bay clerk fell dead. Another shot wounded Governor Semple, and he called to his followers,

"Do what you can to take care of yourselves."

Then there was a volley from the Northwest force, and with the clearing of the smoke it looked as though all the Governor's party were killed or wounded. Instead of taking care of themselves, they had rallied around their wounded leader. Captain Rogers, of the Governor's party, who had fallen, rose to his feet, and ran towards the enemy crying for mercy in English and broken French, when Thomas McKay, a "breed" and Northwest clerk, shot him through the head, another cutting his body open with a knife.

Cuthbert Grant (who, it was charged, had shot Governor Semple) now went to the Governor, while the others despatched the wounded.

Semple said, "Are you not Mr. Grant?"

"Yes," said the other.

"I am not mortally wounded," said the Governor, "and if you could get me conveyed to the fort, I think I should live."

But when Grant left his side an Indian named Ma-chi-ca-taou shot him, some say through the breast, and some have it that he put a pistol to the Governor's head. Grant could not stop the savages. The bloodshed had crazed them. They slaughtered all the wounded, and, worse yet, they terribly maltreated the bodies. Twenty-two Hudson Bay men were killed, and one on the other side was wounded.

There is a story that Alexander McDonnell shouted for joy when he heard the news of the massacre. One witness, who did not hear him shout, reports that he exclaimed to his friends: "*Sacré nom de Dieu! Bonnes nouvelles; vingt-deux Anglais tués!*" (——! Good news; twenty-two English slain!) It was afterwards alleged that the slaughter was approved by every officer of the Northwest Company whose comments were recorded.

It is a saying up in that country that twenty-six out of the sixty-five in the attacking party died violent deaths. The record is only valuable as indicating the nature and perils of the lives the hunters and half-breeds led. First, a Frenchman dropped dead while crossing the ice on the river, his son was stabbed by a comrade, his wife was shot, and his children were burned; "Big Head," his brother, was shot by an Indian; Coutonohais dropped dead at a dance; Battosh was mysteriously shot; Lavigne was drowned; Fraser was run through the body by a Frenchman in Paris; Baptiste Morallé, while drunk, was thrown into a fire by inebriated companions and burned to death; another died drunk on a roadway; another was wounded by the bursting of his gun;

small-pox took the eleventh; Duplicis was empaled upon a hay-fork, on which he jumped from a haystack; Parisien was shot, by a person unknown, in a buffalo-hunt; another lost his arm by carelessness; Gardapie, "the brave," was scalped and shot by the Sioux; so was Vallée; Ka-te-tee-goose was scalped and cut in pieces by the Gros-Ventres; Pe-me-can-toss was thrown in a hole by his people; and another Indian and his wife and children were killed by lightning. Yet another was gored to death by a buffalo. The rest of the twenty-six died by being frozen, by drowning, by drunkenness, or by shameful disease.

It is when things are at their worst that they begin to mend, says a silly old proverb; but when history is studied these desperate situations often seem part of the mending, not of themselves, but of the broken cause of progress. There was a little halt here in Canada, as we shall see, but the seed of settlement had been planted, and thenceforth continued to grow. Lord Selkirk came with all speed, reaching Canada in 1817. It was now an English colony, and when he asked for a body-guard, the Government gave him two sergeants and twelve soldiers of the Régiment de Meuron. He made these the nucleus of a considerable force of Swiss and Germans who had formerly served in that regiment, and he pursued a triumphal progress to what he called his territory of Assiniboin, capturing all the Northwest Company's forts on the route, imprisoning the officers, and sending to jail in Canada all the accessaries to the massacre, on charges of arson, murder, robbery, and "high

A HUDSON BAY MAN (QUAKTER-FREED)

misdemeanors." Such was the prejudice against the Hudson Bay Company and the regard for the home corporation that nearly all were acquitted, and suits for very heavy damages were lodged against him.

Selkirk sought to treat with the Indians for his land, which they said belonged to the Chippeways and the Crees. Five chiefs were found whose right to treat was acknowledged by all. On July 18, 1817, they deeded the territory to the King, "for the benefit of Lord Selkirk," giving him a strip two miles wide on either side of the Red River from Lake

Winnipeg to Red Lake, north of the United States
boundary, and along the Assiniboin from Fort Garry
to the Muskrat River, as well as within two circles
of six miles radius around Fort Garry and Pembina,
now in Dakota. Indians do not know what miles
are; they measure distance by the movement of the
sun while on a journey. They determined two miles
in this case to be "as far as you can see daylight
under a horse's belly on the level prairie." On
account of Selkirk's liberality they dubbed him "the
silver chief." He agreed to give them for the land
200 pounds of tobacco a year. He named his settlement Kildonan, after that place in Helmsdale, Sutherlandshire, Scotland. He died in 1821, and in 1836
the Hudson Bay Company bought the land back
from his heirs for £84,000. The Swiss and Germans of his regiment remained, and many retired
servants of the company bought and settled there,
forming the aristocracy of the place—a queer aristocracy to our minds, for many of the women were
Indian squaws, and the children were "breeds."

Through the perseverance and tact of the Right
Hon. Edward Ellice, to whom the Government had
appealed, all differences between the two great fur-
trading companies were adjusted, and in 1821 a
coalition was formed. At Ellice's suggestion the
giant combination then got from Parliament exclusive privileges beyond the waters that flow into
Hudson Bay, over the Rocky Mountains and to the
Pacific, for a term of twenty years. These extra
privileges were surrendered in 1838, and were renewed for twenty-one years longer, to be revoked, so

far as British Columbia (then New Caledonia) was concerned, in 1858. That territory then became a crown colony, and it and Vancouver Island, which had taken on a colonial character at the time of the California gold fever (1849), were united in 1866. The extra privileges of the fur-traders were therefore not again renewed. In 1868, after the establishment of the Canadian union, whatever presumptive rights the Hudson Bay Company got under Charles II.'s charter were vacated in consideration of a payment by Canada of $1,500,000 cash, one-twentieth of all surveyed lands within the fertile belt, and 50,000 acres surrounding the company's posts. It is estimated that the land grant amounts to 7,000,000 of acres, worth $20,000,000, exclusive of all town sites.

Thus we reach the present condition of the company, more than 220 years old, maintaining 200 central posts and unnumbered dependent ones, and trading in Labrador on the Atlantic; at Massett, on Queen Charlotte Island, in the Pacific; and deep within the Arctic Circle in the north. The company was newly capitalized not long ago with 100,000 shares at £20 ($10,000,000), but, in addition to its dividends, it has paid back £7 in every £20, reducing its capital to £1,300,000. The stock, however, is quoted at its original value. The supreme control of the company is vested in a governor, deputy governor, and five directors, elected by the stockholders in London. They delegate their powers to an executive resident in this country, who was until lately called the "Governor of Rupert's Land," but now is styled the chief commissioner, and is in absolute charge of the company and all its operations. His term of office is unlimited. The present head of the corporation, or governor, is Sir Donald A. Smith, one of the foremost spirits in Canada, who worked his way up from a clerkship in the company. The business of the company is managed on the outfit system, the most old-fogyish, yet by its officers declared to be the most perfect, plan in use by any corporation. The method is to charge against each post all the supplies that are sent to it between June 1st and June 1st each year, and then to set against this the product of each post in furs and in cash received. It used to take seven years to arrive at the figures for a given year, but, owing to improved means of transportation, this is now done in two years.

Almost wherever you go in the newly settled parts of the Hudson Bay territory you find at least one free-trader's shop set up in rivalry with the old company's post. These are sometimes mere storehouses for the furs, and sometimes they look like, and are partly, general country stores. There can be no doubt that this rivalry is very detrimental to the fur trade from the stand-point of the future. The great company can afford to miss a dividend, and can lose at some points while gaining at others, but the free-traders must profit in every district. The consequence is such a reckless destruction of game that the plan adopted by us for our seal-fisheries—the leasehold system—is envied and advocated in Canada. A greater proportion of trapping and an utter unconcern for the destruction of the game at all ages

are now ravaging the wilderness. Many districts
return as many furs as they ever yielded, but the
quantity is kept up at fearful cost by the extermina-
tion of the game. On the other hand, the fortified
wall of posts that opposed the development of Canada, and sent the surplus population of Europe to the
United States, is rid of its palisades and field-pieces,
and the main strongholds of the ancient company
and its rivals have become cities. The old fort on
Vancouver Island is now Victoria; Fort Edmonton
is the seat of law and commerce in the Peace River
region; old Fort William has seen Port Arthur rise
by its side; Fort Garry is Winnipeg; Calgary, the
chief city of Alberta, is on the site of another fort;
and Sault Ste. Marie was once a Northwest post.

But civilization is still so far off from most of the
"factories," as the company's posts are called, that
the day when they shall become cities is in no man's
thought or ken. And the communication between
the centres and outposts is, like the life of the traders,
more nearly like what it was in the old, old days than
most of my readers would imagine. My Indian
guides were battling with their paddles against the
mad current of the Nipigon, above Lake Superior,
one day last summer, and I was only a few hours
away from Factor Flanagan's post near the great
lake, when we came to a portage, and might have
imagined from what we saw that time had pushed the
hands back on the dial of eternity at least a century.

Some rapids in the river had to be avoided by the
brigade that was being sent with supplies to a post
far north at the head of Lake Nipigon. A cumbrous,

big-timbered little schooner, like a surf-boat with a sail, and a square-cut bateau had brought the men and goods to the "carry." The men were half-breeds as of old, and had brought along their women and children to inhabit a camp of smoky tents that we espied on a bluff close by; a typical camp, with the blankets hung on the bushes, the slatternly women and half-naked children squatting or running about, and smudge fires smoking between the tents to drive off mosquitoes and flies. The men were in groups below on the trail, at the water-side end of which were the boats' cargoes of shingles and flour and bacon and shot and powder in kegs, wrapped, two at a time, in rawhide. They were dark-skinned, short, spare men, without a surplus pound of flesh in the crew, and with longish coarse black hair and straggling beards. Each man carried a tump-line, or long stout strap, which he tied in such a way around what he meant to carry that a broad part of the strap fitted over the crown of his head. Thus they "packed" the goods over the portage, their heads sustaining the loads, and their backs merely steadying them. When one had thrown his burden into place, he trotted off up the trail with springing feet, though the freight was packed so that 100 pounds should form a load. For bravado one carried 200 pounds, and then all the others tried to pack as much, and most succeeded. All agreed that one, the smallest and least muscular-looking one among them, could pack 400 pounds.

As the men gathered around their "smudge" to talk with my party, it was seen that of all the parts of the picturesque costume of the *voyageur* or *bois-brûlé* of old—the capote, the striped shirt, the pipe-tomahawk, plumed hat, gay leggins, belt, and moccasins—only the red worsted belt and the moccasins have been retained. These men could recall the day when they had tallow and corn meal for rations, got no tents, and were obliged to carry 200 pounds, lifting one package, and then throwing a second one atop of it without assistance. Now they carry only 100 pounds at a time, and have tents and good food given to them.

We will not follow them, nor meet, as they did, the York boat coming down from the north with last winter's furs. Instead, I will endeavor to lift the curtain from before the great fur country beyond them, to give a glimpse of the habits and conditions that prevail throughout a majestic territory where the rivers and lakes are the only roads, and canoes and dog-sleds are the only vehicles.

VII

"TALKING MUSQUASH"

Concluding the sketch of the history and work of the Hudson Bay Company

THE most sensational bit of "musquash talk" in more than a quarter of a century among the Hudson Bay Company's employés was started the other day, when Sir Donald A. Smith, the governor of the great trading company, sent a type-written letter to Winnipeg. If a Cree squaw had gone to the trading-shop at Moose Factory and asked for a bustle and a box of face-powder in exchange for a beaver-skin, the suggestion of changing conditions in the fur trade would have been trifling compared with the sense of instability to which this appearance of machine-writing gave rise. The reader may imagine for himself what a wrench civilization would have gotten if the world had laid down its goose-quills and taken up the typewriter all in one day. And that is precisely what Sir Donald Smith had done. The quill that had served to convey the orders of Alexander Mackenzie had satisfied Sir George Simpson; and, in our own time, while men like Lord Iddesleigh, Lord Kimberley, and Mr. Goschen sat around the candle-lighted table in the board-room of the company in London, quill pens were the only ones at hand. But Sir Donald's letter was not only the product of a machine; it contained instructions for the use of the type-writer in the offices at Winnipeg, and there was in the letter a protest against illegible manual chirography such as had been received from many factories in the wilderness. Talking business in the fur trade has always been called "talking musquash" (musk-rat), and after that letter came the turn taken by that form of talk suggested a general fear that from the Arctic to our border and from Labrador to Queen Charlotte's Islands the canvassers for competing machines will be "racing" in all the posts, each to prove that his instrument can pound out more words in a minute than any other—in those posts where life has hitherto been taken so gently that when one day a factor heard that the battle of Waterloo had been fought and won by the English, he deliberately loaded the best trade gun in the storehouse and went out and fired it into the pulseless woods, although it was two years after the battle, and the disquieted Old World had long known the greater news that Napoleon was caged in St. Helena. The only reassuring note in the "musquash talk" to-day is sounded when the subject of candles is reached. The Governor and committee in London still pursue their deliberations by candlelight.

But rebellion against their fate is idle, and it is of no avail for the old factors to make the point that Sir Donald found no greater trouble in reading their writing than they encountered when one of his missives had to be deciphered by them. The truth is that the tide of immigration which their ancient monopoly first shunted into the United States is now sweeping over their vast territory, and altering more than its face. Not only are the factors aware that the new rule confining them to share in the profits of the fur trade leaves to the mere stockholders far greater returns from land sales and storekeeping, but a great many of them now find village life around their old forts, and railroads close at hand, and Law setting up its officers at their doors, so that in a great part of the territory the romance of the old life, and their authority as well, has fled.

Less than four years ago I had passed by Qu'Appelle without visiting it, but last summer I resolved not to make the mistake again, for it was the last stockaded fort that could be studied without a tiresome and costly journey into the far north. It is on the Fishing Lakes, just beyond Manitoba. But on my way a Hudson Bay officer told me that they had just taken down the stockade in the spring, and that he did not know of a remaining "palisadoe" in all the company's system except one, which, curiously enough, had just been ordered to be put up around Fort Hazleton, on the Skeena River, in northern British Columbia, where some turbulent Indians have been very troublesome, and where whatever civilization there may be in Saturn seems nearer than our own. This one example of the survival of original conditions is far more eloquent of their endurance than the thoughtless reader would imagine. It is true that there has come a tremendous change in the status and spirit of the company. It is true that its officers are but newly bending to external authority, and that settlers have poured into the

TALKING MUSQUASH

south with such demands for food, clothes, tools, and weapons as to create within the old corporation one of the largest of shopkeeping companies. Yet to-day, as two centuries ago, the Hudson Bay Company remains the greatest fur-trading association that exists.

The zone in which Fort Hazleton is situated reaches from ocean to ocean without suffering invasion by settlers, and far above it to the Arctic Sea is a grand belt wherein time has made no impress since the first factory was put up there. There and around it is a region, nearly two-thirds the size of the United States, which is as if our country were meagrely dotted with tiny villages at an average distance of five days apart, with no other means of communication than canoe or dog train, and with not above a thousand white men in it, and not as many pure-blooded white women as you will find registered at a first-class New York hotel on an ordinary day. The company employs between fifteen hundred and two thousand white men, and I am assuming that half of them are in the fur country.

We know that for nearly a century the company clung to the shores of Hudson Bay. It will be interesting to peep into one of its forts as they were at that time; it will be amazing to see what a country that bay-shore territory was and is. There and over a vast territory three seasons come in four months— spring in June, summer in July and August, and autumn in September. During the long winter the earth is blanketed deep in snow, and the water is locked beneath ice. Geese, ducks, and smaller birds abound as probably they are not seen elsewhere in America, but they either give place to or share the summer with mosquitoes, black-flies, and "bull-dogs" (*tabanus*) without number, rest, or mercy. For the land around Hudson Bay is a vast level marsh, so wet that York Fort was built on piles, with elevated platforms around the buildings for the men to walk upon. Infrequent bunches of small pines and a litter of stunted swamp-willows dot the level waste, the only considerable timber being found upon the banks of the rivers. There is a wide belt called the Arctic Barrens all along the north, but below that, at some distance west of the bay, the great forests of Canada bridge across the region north of the prairie and the plains, and cross the Rocky Mountains to reach the Pacific. In the far north the musk-ox descends almost to meet the moose and deer, and on the near slope of the Rockies the wood-buffalo—larger, darker, and fiercer than the bison of the plains, but very like him—still roams as far south as where the buffalo ran highest in the days when he existed.

Through all this northern country the cold in winter registers 40°, and even 50°, below zero, and the travel is by dogs and sleds. There men in camp may be said to dress to go to bed. They leave their winter's store of dried meat and frozen fish out-of-doors on racks all winter (and so they do down close to Lake Superior); they hear from civilization only twice a year at the utmost; and when supplies have run out at the posts, we have heard of their boiling the parchment sheets they use instead of glass in their windows, and of their cooking the fat out of

beaver-skins to keep from starving, though beaver is
so precious that such recourse could only be had
when the horses and dogs had been eaten. As to
the value of the beaver, the reader who never has
purchased any for his wife may judge what it must
be by knowing that the company has long imported
buckskin from Labrador to sell to the Chippeways
around Lake Nipigon in order that they may not be
tempted, as of old, to make thongs and moccasins of
the beaver; for their deer are poor, with skins full of
worm-holes, whereas beaver leather is very tough and
fine.

But in spite of the severe cold winters, that are, in
fact, common to all the fur territory, winter is the delightful season for the traders; around the bay it is
the only endurable season. The winged pests of
which I have spoken are by no means confined to
the tide-soaked region close to the great inland sea.
The whole country is as wet as that orange of which
geographers speak when they tell us that the water
on the earth's surface is proportioned as if we were
to rub a rough orange with a wet cloth. Up in
what we used to call British America the illustra-
tion is itself illustrated in the countless lakes of all
sizes, the innumerable small streams, and the many
great rivers that make waterways the roads, as canoes
are the wagons, of the region. It is a vast paradise
for mosquitoes, and I have been hunted out of fishing and hunting grounds by them as far south as the
border. The "bull-dog" is a terror reserved for especial districts. He is the Sioux of the insect world,
as pretty as a warrior in buckskin and beads, but car-

rying a red-hot sword blade, which, when sheathed in human flesh, will make the victim jump a foot from the ground, though there is no after-pain or itching or swelling from the thrust.

Having seen the country, let us turn to the forts. Some of them really were forts, in so far as palisades and sentry towers and double doors and guns can make a fort, and one twenty

INDIAN HUNTERS MOVING CAMP

miles below Winnipeg was a stone fort. It is still standing. When the company ruled the territory as its landlord, the defended posts were on the plains among the bad Indians, and on the Hudson Bay shore, where vessels of foreign nations might be expected. In the forests, on the lakes and rivers, the character and behavior of the fish-eating Indians did not warrant armament. The stockaded forts were nearly all alike. The stockade was of timber, of about such a height that a man might look over it on tiptoe. It had towers at the corners, and York Fort had a great "lookout" tower within the enclosure. Within the barricade were the company's buildings, making altogether such a picture as New York presented when the Dutch founded it and called it New Amsterdam, except that we had a church and a stadt-house in our enclosure. The Hudson Bay buildings were sometimes arranged in a hollow square, and sometimes in the shape of a letter H, with the factor's house connecting the two other parts of the character. The factor's house was the best dwelling, but there were many smaller ones for the laborers, mechanics, hunters, and other non-commissioned men. A long, low, whitewashed log-house was apt to be the clerks' house, and other large buildings were the stores where merchandise was kept, the fur-houses where the furs, skins, and pelts were stored, and the Indian trading-house, in which all the bartering was done. A powder-house, ice-house, oil-house, and either a stable or a boat-house for canoes completed the post. All the houses had double doors and windows, and wherever the men lived there was a tremendous stove set up to battle with the cold.

The abode of jollity was the clerks' house, or bachelors' quarters. Each man had a little bedroom containing his chest, a chair, and a bed, with the walls covered with pictures cut from illustrated papers or not, according to each man's taste. The big room or hall, where all met in the long nights and on off days, was as bare as a baldpate so far as its whitewashed or timbered walls went, but the table in the middle was littered with pipes, tobacco, papers, books, and pens and ink, and all around stood (or rested on hooks overhead) guns, foils, and fishing-rods. On Wednesdays and Saturdays there was no work in at least one big factory. Breakfast was served at nine o'clock, dinner at one o'clock, and tea at six o'clock. The food varied in different places. All over the prairie and plains great stores of pemmican were kept, and men grew to like it very much, though it was nothing but dried buffalo beef pounded and mixed with melted fat. But where they had pemmican they also enjoyed buffalo hunch in the season, and that was the greatest delicacy, except moose muffle (the nose of the moose), in all the territory. In the woods and lake country there were venison and moose as well as beaver—which is very good eating—and many sorts of birds, but in that region dried fish (salmon in the west, and lake trout or white-fish nearer the bay) was the staple. The young fellows hunted and fished and smoked and drank and listened to the songs of the *voyageurs* and the yarns of the "breeds" and Indians. For the rest there was plenty of work to do.

They had a costume of their own, and, indeed, in that respect there has been a sad change, for all the

SETTING A MINK-TRAP

people, white, red, and crossed, dressed picturesquely. You could always distinguish a Hudson Bay man by his capote of light blue cloth with brass buttons. In winter they wore as much as a Quebec carter. They wore leather coats lined with flannel, edged with fur, and double-breasted. A scarlet worsted belt went around their waists, their breeches were of smoked buckskin, reaching down to three pairs of blanket socks and moose moccasins, with blue cloth leggins up to the knee. Their buckskin mittens were hung from their necks by a cord, and usually they wrapped a shawl of Scotch plaid around their necks and shoulders, while on each one's head was a fur cap with ear-pieces.

The French Canadians and "breeds," who were the *voyageurs* and hunters, made a gay appearance. They used to wear the company's regulation light blue capotes, or coats, in winter, with flannel shirts, either red or blue, and corduroy trousers gartered at the knee with bead-work. They all wore gaudy worsted belts, long, heavy woollen stockings — covered with gayly-fringed leggins — fancy moccasins, and tuques, or feather-decked hats or caps bound with tinsel bands. In mild weather their costume was formed of a blue striped cotton shirt, corduroys, blue cloth leggins bound with orange ribbons, the inevitable sash or worsted belt, and moccasins. Every hunter carried a powder-horn slung from his neck, and in his belt a tomahawk, which often served also as a pipe. As late as 1862, Viscount Milton and W. B. Cheadle describe them in a book, *The Northwest Passage by Land*, in the following graphic language:

"The men appeared in gaudy array, with beaded fire-bag, gay sash, blue or scarlet leggings, girt below the knee with beaded garters, and moccasins elaborately embroidered. The (half-breed) women were in short, bright-colored skirts, showing richly embroidered leggings and white moccasins of cariboo-skin beautifully worked with flowery patterns in beads, silk, and moose hair."

The trading-room at an open post was — and is now — like a cross-roads store, having its shelves laden with every imaginable article that Indians like and hunters need—clothes, blankets, files, scalp-knives, gun screws, flints, twine, fire-steels, awls, beads, needles, scissors, knives, pins, kitchen ware, guns, powder, and shot. An Indian who came in with furs threw them down, and when they were counted received the right number of castors—little pieces of wood which served as money—with which, after the hours of reflection an Indian spends at such a time, he bought what he wanted.

But there was a wide difference between such a trading-room and one in the plains country, or where there were dangerous Indians—such as some of the Crees, and the Chippeways, Blackfeet, Bloods, Sarcis, Sioux, Sicanies, Stonies, and others. In such places the Indians were let in only one or two at a time, the goods were hidden so as not to excite their cupidity, and through a square hole grated with a cross of iron, whose spaces were only large enough to pass a blanket, what they wanted was given to them. That is all done away with now, except it be in northern British Columbia, where the Indians have been turbulent.

Farther on we shall perhaps see a band of Indians on their way to trade at a post. Their custom is to wait until the first signs of spring, and then to pack up their winter's store of furs, and take advantage of the last of the snow and ice for the journey. They hunt from November to May; but the trapping and shooting of bears go on until the 15th of June, for those animals do not come from their winter dens until May begins. They come to the posts in their best attire, and in the old days that formed as strong a contrast to their present dress as their leather tepees of old did to the cotton ones of to-day. Ballantyne, who wrote a book about his service with the great fur company, says merely that they were painted, and with scalp-locks fringing their clothes; but in Lewis and Clarke's journal we read description after description of the brave costuming of these color-and-ornament-loving people. Take the Sioux, for instance. Their heads were shaved of all but a tuft of hair, and feathers hung from that. Instead of the universal blanket of to-day, their main garment was a robe of buffalo-skin with the fur left on, and the inner surface dressed white, painted gaudily with figures of beasts and queer designs, and fringed with porcupine quills. They wore the fur side out only in wet weather. Beneath the robe they wore a shirt of dressed skin, and under that a leather belt, under which the ends of a breech-clout of cloth, blanket stuff, or skin were tucked. They wore leggins of dressed antelope hide with scalp-locks fringing the seams, and prettily beaded moccasins for their feet. They had necklaces of the teeth or claws of wild beasts, and each carried a fire-bag, a quiver, and a brightly painted shield, giving up the quiver and shield when guns came into use.

The Indians who came to trade were admitted to the store precisely as voters are to the polls under the Australian system—one by one. They had to leave their guns outside. When rum was given out, each Indian had to surrender his knife before he got his tin cup.

The company made great use of the Iroquois, and considered them the best boatmen in Canada. Sir Alexander Mackenzie, of the Northwest Company, employed eight of them to paddle him to the Pacific Ocean by way of the Peace and Fraser rivers, and

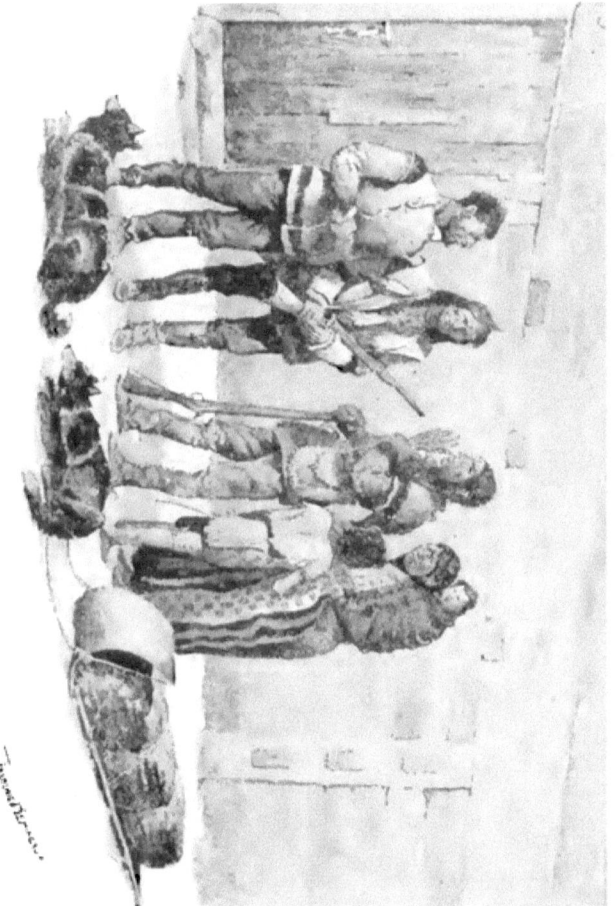

WOOD INDIANS COME TO TRADE

when the greatest of Hudson Bay executives, Sir George Simpson, travelled, Iroquois always propelled him. The company had a uniform for all its Indian employés—a blue, gray, or blanket capote, very loose, and reaching below the knee, with a red worsted belt around the waist, a cotton shirt, no trousers, but artfully beaded leggins with wide flaps at the seams, and moccasins over blanket socks. In winter they wore buckskin coats lined with flannel, and mittens were given to them. We have seen how the half-breeds were dressed. They were long employed at women's work in the forts, at making clothing and at mending. All the mittens, moccasins, fur caps, deerskin coats, etc., were made by them. They were also the washer-women.

Perhaps the factor had a good time in the old days, or thought he did. He had a wife and servants and babies, and when a visitor came, which was not as often as snow-drifts blew over the stockade, he entertained like a lord. At first the factors used to send to London, to the head office, for a wife, to be added to the annual consignment of goods, and there must have been a few who sent to the Orkneys for the sweethearts they left there. But in time the rule came to be that they married Indian squaws. In doing this, not even the first among them acted blindly, for their old rivals and subsequent companions of the Northwest and X. Y. companies began the custom, and the French *voyageurs* and *coureurs du bois* had mated with Indian women before there was a Hudson Bay Company. These rough and hardy woodsmen, and a large number of half-breeds born of just such alliances, began at an early day to settle near the trading-posts. Sometimes they established what might be called villages, but were really close imitations of Indian camps, composed of a cluster of skin tepees, racks of fish or meat, and a swarm of dogs, women, and children. In each tepee was the fireplace, beneath the flue formed by the open top of the habitation, and around it were the beds of brush, covered with soft hides, the inevitable copper kettle, the babies swaddled in blankets or moss bags, the women and dogs, the gun and paddle, and the junks and strips of raw meat hanging overhead in the smoke. This has not changed to-day; indeed, very little that I shall speak of has altered in the true or far fur country. The camps exist yet. They are not so clean (or, rather, they are more dirty), and the clothes and food are poorer and harder to get; that is all.

The Europeans saw that these women were docile, or were kept in order easily by floggings with the tent poles; that they were faithful and industrious, as a rule, and that they were not all unprepossessing —from their point of view, of course. Therefore it came to pass that these were the most frequent alliances in and out of the posts in all that country. The consequences of this custom were so peculiar and important that I must ask leave to pause and consider them. In Canada we see that the white man thus made his bow to the redskin as a brother in the truest sense. The old *coureurs* of Norman and Breton stock, loving a wild, free life, and in complete sympathy with the Indian, bought or took the squaws to wife, learned the Indian dialects, and shared their food and adventures with the tribes. As more and more entered the wilderness, and at last came to be supported, in camps and at posts and as *voyageurs*, by the competing fur companies, there grew up a class of half-breeds who spoke English and French, married Indians, and were as much at home with the savages as with the whites. From this stock the Hudson Bay men have had a better choice of wives for more than a century. But when these "breeds" were turbulent and murderous—first in the attacks on Selkirk's colony, and next during the Riel rebellion—the Indians remained quiet. They defined their position when, in 1819, they were tempted with great bribes to massacre the Red River colonists.

A VOYAGEUR OR CANOE-MAN OF GREAT SLAVE LAKE.

"No," said they; "the colonists are our friends." The men who sought to excite them to murder were the officers of the Northwest Company, who bought furs of them, to be sure, but the colonists had shared with the Indians in poverty and plenty, giving now and taking then. All were alike to the red men—friends, white men, and of the race that had taken so many of their women to wife. Therefore they went to the colonists to tell them what was being planned against them, and not from that day to this has an Indian band taken the war-path against the Canadians. I have read General Custer's theory that the United States had to do with meat-eating Indians, whereas the Canadian tribes are largely fish-eaters, and I have seen 10,000 references to the better Indian policy of Canada; but I can see no difference in the two policies, and between the Rockies and the Great Lakes I find that Canada had the Stonies, Blackfeet, and many other fierce tribes of buffalo-hunters. It is in the slow, close-growing acquaintance between the two races, and in the just policy of the Hudson Bay men towards the Indians, that I see the reason for Canada's enviable experience with her red men.

But even the Hudson Bay men have had trouble with the Indians in recent years, and one serious affair grew out of the relations between the company's servants and the squaws. There is etiquette even among savages, and this was ignored up at old Fort St. Johns, on the Peace River, with the result that the Indians slaughtered the people there and burned the fort. They were Sicanie Indians of that region, and after they had massacred the men in charge, they met a boat-load of white men coming up the river with goods. To them they turned their guns also, and only four escaped. It was up in that

IN A STIFF CURRENT

country likewise—just this side of the Rocky Mountains, where the plains begin to be forested—that a silly clerk in a post quarrelled with an Indian, and said to him, "Before you come back to this post again, your wife and child will be dead." He spoke hastily, and meant nothing, but squaw and pappoose happened to die that winter, and the Indian walked into the fort the next spring and shot the clerk without a word.

To-day the posts are little village-like collections of buildings, usually showing white against a green background in the prettiest way imaginable; for, as a rule, they cluster on the lower bank of a river, or the lower near shore of a lake. There are not clerks enough in most of them to render a clerks' house necessary, for at the little posts half-breeds are seen to do as good service as Europeans. As a rule, there is now a store or trading-house and a fur-house and the factor's house, the canoe-house and the stable, with a barn where gardening is done, as is often the case when soil and climate permit. Often the fur-house and store are combined, the furs being laid in the upper story over the shop. There is always a flag-staff, of course. This and the flag, with the letters "H. B. C." on its field, led to the old hunters' saying that the initials stood for "Here before Christ," because, no matter how far away from the frontier a man might go, in regions he fancied no white man had been, that flag and those letters stared him in the face. You will often find that the factor, rid of all the ancient timidity that called for "palisadoes and swivels," lives on the high upper bank above the store. The usual half-breed or Indian village is seldom farther than a couple of miles away, on the same water. The factor is still, as he always has been, responsible only to himself for the discipline and management of his post, and therefore among the factories we will find all sorts of homes—homes where a piano and the magazines are prized, and daughters educated abroad shed the lustre of refinement upon their surroundings, homes where no woman rules, and homes of the French half-breed type, which we shall see is a very different mould from that of the two sorts of British half-breed that are numerous. There never was a rule by which to gauge a post. In one you found religion valued and missionaries welcomed, while in others there never was sermon or hymn. In some, Hudson Bay rum met the rum of the free-traders, and in others no rum was bartered away. To-day, in this latter respect, the Dominion law prevails, and rum may not be given or sold to the red man.

When one thinks of the lives of these factors, hidden away in forest, mountain chain, or plain, or arctic barren, seeing the same very few faces year in and year out, with breaches of the monotonous routine once a year when the winter's furs are brought in, and once a year when the mail-packet arrives—when one thinks of their isolation, and lack of most of those influences which we in our walks prize the highest, the reason for their choosing that company's service seems almost mysterious. Yet they will tell you there is a fascination in it. This could be understood so far as the half-breeds and French Canadians were concerned, for they inherited the liking; and, after all, though most of them are only laborers, no other laborers are so free, and none spice life with so much of adventure. But the factors are mainly men of ability and good origin, well fitted to occupy responsible positions, and at better salaries. However, from the outset the rule has been that they have become as enamoured of the trader's life as soldiers and sailors always have of theirs. They have usually retired from it reluctantly, and some, having gone home to Europe, have begged leave to return.

The company has always been managed upon something like a military basis. Perhaps the original necessity for forts and men trained to the use of arms suggested this. The uniforms were in keeping with the rest. The lowest rank in the service is that of the laborer, who may happen to fish or hunt at times, but is employed—or enlisted, as the fact is, for a term of years—to cut wood, shovel snow, act as a porter or gardener, and labor generally about the post. The interpreter was usually a promoted laborer, but long ago the men in the trade, Indians and whites alike, met each other half-way in the matter of language. The highest non-commissioned rank in early days was that of the postmaster at large posts. Men of that rank often got charge of small outposts, and we read that they were "on terms of equality with gentlemen." To-day the service has lost these fine points, and the laborers and commissioned officers are sharply separated. The so-called "gentleman" begins as a prentice clerk, and after a few years becomes a clerk. His next elevation is to the rank of a junior chief trader, and so on through the grades of chief trader, factor, and chief factor, to the office of chief commissioner, or resident American manager, chosen by the London board, and having full powers delegated to him. A clerk—or "clark," as the rank is called—may never touch a pen. He may be a trader. Then again he may be truly an accountant. With the rank he gets a commission, and that entitles him to a minimum guarantee, with a conditional extra income based on the profits of the fur trade. Men get promotions through the chief commissioner, and he has always made fitness, rather than seniority, the criterion. Retiring officers are salaried for a term of years, the original pension fund and system having been broken up.

Sir Donald A. Smith, the present governor of the company, made his way to the highest post from the place of a prentice clerk. He came from Scotland as a youth, and after a time was so unfortunate as to be sent to the coast of Labrador, where a man is as much out of both the world and contact with the heart of the company as it is possible to be. The military system was felt in that instance; but every man who accepts a commission engages to hold himself in readiness to go cheerfully to the north pole, or anywhere between Labrador and the Queen Charlotte Islands. However, to a man of Sir Donald's parts no obstacle is more than a temporary impediment. Though he stayed something like seventeen years in Labrador, he worked faithfully when there was work to do, and in his own time he read and studied voraciously. When the Riel rebellion—the first one — disturbed the country's peace, he appeared on the scene as commissioner for the Government. Next he became chief commissioner for the Hudson Bay Company. After a time he resigned that office to go on the board in London, and thence he stepped easily to the governorship. His parents, whose home was in Morayshire, Scotland, gave him at his birth, in 1821, not only a constitution of iron, but that shrewdness which is only

VOYAGEUR WITH TUMPLINE.

Scotch, and he afterwards developed remarkable foresight, and such a grasp of affairs and of complex situations as to amaze his associates.

Of course his career is almost as singular as his gifts, and the governorship can scarcely be said to be the goal of the general ambition, for it has been most apt to go to a London man. Even ordinary promotion in the company is very slow, and it follows that most men live out their existence between the rank of clerk and that of chief factor. There are 200 central posts, and innumerable dependent posts, and the officers are continually travelling from one to another, some in their districts, and the chief or supervising ones over vast reaches of country. In winter, when dogs and sleds are used, the men walk, as a rule, and it has been nothing for a man to trudge 1000 miles in that way on a winter's journey. Roderick Macfarlane, who was cut off from the world up in the Mackenzie district, became an indefatigable explorer, and made most of his journeys on snow-shoes. He explored the Peel, the Liard, and the Mackenzie, and their surrounding regions, and went far within the Arctic Circle, where he founded the most northerly post of the company. By the regular packet from Calgary, near our border, to the northernmost post is a 3000-mile journey. Macfarlane was fond of the study of ornithology, and classified and catalogued all the birds that reach the frozen regions.

I heard of a factor far up on the east side of Hudson Bay who reads his daily newspaper every morning with his coffee—but of course such an instance is a rare one. He manages it by having a complete set of the London *Times* sent to him by each winter's packet, and each morning the paper of that date in the preceding year is taken from the bundle by his servant and dampened, as it had been when it left the press, and spread by the factor's plate. Thus he gets for half an hour each day a taste of his old habit and life at home.

There was another factor who developed artistic capacity, and spent his leisure at drawing and painting. He did so well that he ventured many sketches for the illustrated papers of London, some of which were published.

The half-breed has developed with the age and growth of Canada. There are now half-breeds and half-breeds, and some of them are titled, and others hold high official places. It occurred to an English lord not long ago, while he was being entertained in a Government house in one of the parts of newer Canada, to inquire of his host, "What are these halfbreeds I hear about? I should like to see what one looks like." His host took the nobleman's breath away by his reply. "I am one," said he. There is no one who has travelled much in western Canada who has not now and then been entertained in homes where either the man or woman of the household was of mixed blood, and in such homes I have found a high degree of refinement and the most polished manners. Usually one needs the information that such persons possess such blood. After that the peculiar black hair and certain facial features in the subject of such gossip attest the truthfulness of the assertion. There is no rule for measuring the character and quality of this plastic, receptive, and often very ambitious element in Canadian society, yet one may say broadly that the social position and attainments of these people have been greatly influenced by the nationality of their fathers. For instance, the French *habitants* and woodsmen far, far too often sank to the level of their wives when they married Indian women. Light-hearted, careless, unambitious, and drifting to the wilderness because of the absence of restraint there; illiterate, of coarse origin, fond of whiskey and gambling—they threw off superiority to the Indian, and evaded responsibility and concern in home management. Of course this is not a rule, but a tendency. On the other hand, the Scotch and English forced their wives up to their own standards. Their own home training, respect for more than the forms of religion, their love of home and of a permanent patch of ground of their own—all these had their effect, and that has been to rear half-breed children in proud and comfortable homes, to send them to mix with the children of cultivated persons in old communities, and to fit them with pride and ambition and cultivation for an equal start in the journey of life. Possessing such foundation for it, the equality has happily never been denied to them in Canada.

To-day the service is very little more inviting than in the olden time. The loneliness and removal from the touch of civilization remain throughout a vast region; the arduous journeys by sled and canoe remain; the dangers of flood and frost are undiminished. Unfortunately, among the changes made by time, one is that which robs the present factor's surroundings of a great part of that which was most picturesque. Of all the prettinesses of the Indian costuming one sees now only a trace here and there in a few tribes, while in many the moccasin and tepee, and in some only the moccasin, remain. The birch-bark canoe and the snow-shoe are the main reliance of both races, but the steamboat has been impressed into parts of the service, and most of the descendants of the old-time *voyageur* preserve only his worsted belt, his knife, and his cap and mocca-

VOYAGEURS IN CAMP FOR THE NIGHT

sins at the utmost. In places the *engagé* has become a mere deck-hand. His scarlet paddle has rotted away; he no longer awakens the echoes of forest or cañon with *chansons* that died in the throats of a generation that has gone. In return, the horrors of intertribal war and of a precarious foothold among fierce and turbulent bands have nearly vanished; but there was a spice in them that added to the fascination of the service.

The dogs and sleds form a very interesting part of the Hudson Bay outfit. One does not need to go very deep into western Canada to meet with them. As close to our centre of population as Nipigon, on Lake Superior, the only roads into the north are the rivers and lakes, traversed by canoes in summer and sleds in winter. The dogs are of a peculiar breed, and are called "huskies"—undoubtedly a corruption of the word Esquimaux. They preserve a closer resemblance to the wolf than any of our domesticated dogs, and exhibit their kinship with that scavenger of the wilderness in their nature as well as their looks. To-day their females, if tied and left in the forest, will often attest companionship with its denizens by bringing forth litters of wolfish progeny. Moreover, it will not be necessary to feed all with whom the experiment is tried, for the wolves will be apt to bring food to them as long as they are thus neglected by man. They are often as large as the ordinary Newfoundland dog, but their legs are shorter, and even more hairy, and the hair along their necks, from their shoulders to their skulls, stands erect in a thick, bristling mass. They have the long snouts, sharp-pointed ears, and the tails of wolves, and their cry is a yelp rather than a bark. Like wolves they are apt to yelp in chorus at sunrise and at sunset. They delight in worrying peaceful animals, setting their own numbers against one, and they will kill cows, or even children, if they get the chance. They are disciplined only when at work, and are then so surprisingly obedient, tractable, and industrious as to plainly show that though their nature is savage and wolfish, they could be reclaimed by domestication. In isolated cases plenty of them are. As it is, in their packs, their battles among themselves are terrible, and they are dangerous when loose. In some districts it is the custom to turn them loose in summer on little islands in the lakes, leaving them to hunger or feast according as the supply of dead fish thrown upon the shore is small or plentiful. When they are kept in dog quarters they are simply penned up and fed during the summer, so that the savage side of their nature gets full play during long periods. Fish is their principal diet, and stores of dried fish are kept for their winter food. Corn meal is often fed to them also. Like a wolf or an Indian, a "husky" gets along without food when there is not any, and will eat his own weight of it when it is plenty.

A typical dog-sled is very like a toboggan. It is formed of two thin pieces of oak or birch lashed together with buckskin thongs and turned up high in front. It is usually about nine feet in length by sixteen inches wide. A leather cord is run along the outer edges for fastening whatever may be put upon the sled. Varying numbers of dogs are harnessed to such sleds, but the usual number is four. Traces, collars, and backbands form the harness, and the dogs are hitched one before the other. Very often the collars are completed with sets of sleigh-bells, and sometimes the harness is otherwise ornamented with beads, tassels, fringes, or ribbons. The leader, or fore-goer, is always the best in the team. The dog next to him is called the steady dog, and the last is named the steer dog. As a rule, these faithful animals are treated harshly, if not brutally. It is a Hudson Bay axiom that no man who cannot curse in three languages is fit to drive them. The three profanities are, of course, English, French, and Indian, though whoever has heard the Northwest French knows that it ought to serve by itself, as it is half-soled with Anglo-Saxon oaths and heeled with Indian obscenity. The rule with whoever goes on a dog-sled journey is that the driver, or mock-passenger, runs behind the dogs. The main function of the sled is to carry the dead weight, the burdens of tent-covers, blankets, food, and the like. The men run along with or behind the dogs, on snow-shoes, and when the dogs make better time than horses are able to, and will carry between 200 and 300 pounds over daily distances of from 20 to 35 miles, according to the condition of the ice or snow, and that many a journey of 1000 miles has been performed in this way, and some of 2000 miles, the test of human endurance is as great as that of canine grit.

Men travelling "light," with extra sleds for the freight, and men on short journeys often ride in the sleds, which in such cases are fitted up as "carioles" for the purpose. I have heard an unauthenticated account, by a Hudson Bay man, of men who drove themselves, disciplining refractory or lazy dogs by simply pulling them in beside or over the dash-board, and holding them down by the neck while they thrashed them. A story is told of a worthy bishop who complained of the slow progress his sled was making, and was told that it was useless to complain, as the dogs would not work unless they were roundly and incessantly cursed. After a time the bishop gave his driver absolution for the profanity needed for the remainder of the journey, and thenceforth sped over the snow at a gallop, every stroke of the half-breed's long and cruel whip being sent home with a volley of wicked words, emphasized at times with peltings with sharp-edged bits of ice. Kane, the explorer, made an average of 57 miles a day behind these shaggy little brutes. Milton and Cheadle, in their book, mention instances where the dogs made 140 miles in less than 48 hours, and the Bishop of Rupert's Land told me he had covered 20 miles in a forenoon and 20 in the afternoon of the same day, without causing his dogs to exhibit evidence of fatigue. The best time is made on hard snow and ice, of course, and when the conditions suit, the drivers whip off their snow-shoes to trot behind the dogs more easily. In view of what they do, it is no wonder that many of the Northern Indians, upon first seeing horses, named them simply "big dog." But to me the performances of the drivers are the more wonderful. It was a white youth, son of a factor,

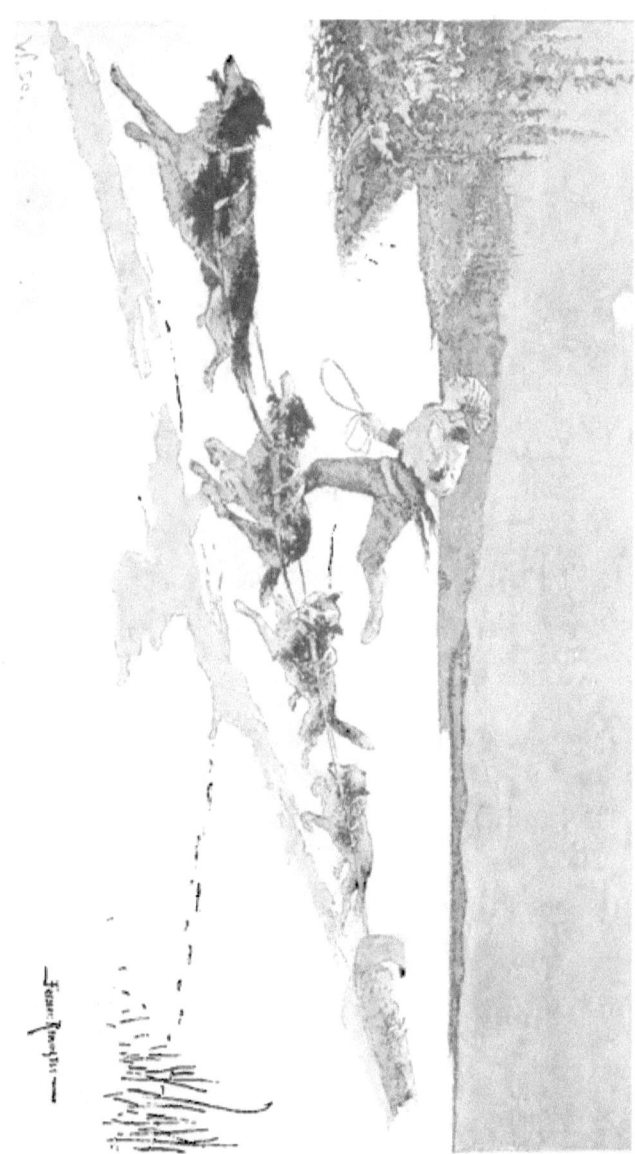

who ran behind the bishop's dogs in the spurt of 40 miles by daylight that I mention. The men who do such work explain that the "lope" of the dogs is peculiarly suited to the dog-trot of a human being.

A picture of a factor on a round of his outposts, or of a chief factor racing through a great district, will now be intelligible. If he is riding, he fancies that princes and lords would envy him could they see his luxurious comfort. Fancy him in a dog-cariole of the best pattern—a little suggestive of a burial casket, to be sure, in its shape, but gaudily painted, and so full of soft warm furs that the man within is enveloped like a chrysalis in a cocoon. Perhaps there are Russian bells on the collars of the dogs, and their harness is "Frenchified" with bead-work and tassels. The air, which fans only his face, is crisp and invigorating, and before him the lake or stream over which he rides is a sheet of virgin snow—not nature's winding-sheet, as those who cannot love nature have said, but rather a robe of beautiful ermine fringed and embroidered with dark evergreen, and that in turn flecked at every point with snow, as if bejewelled with pearls. If the factor chats with his driver, who falls behind at rough places to keep the sled from tipping over, their conversation is carried on at so high a tone as to startle the birds into flight, if there are any, and to shock the scene as by the greatest rudeness possible in that then vast, silent land. If silence is kept, the factor reads the prints of game in the snow, of foxes' pads and deer hoofs, of wolf splotches, and the queer hieroglyphics of birds, or the dots and troughs of rabbit-trailing. To him these are as legible as the Morse alphabet to telegraphers, and as important as stock quotations to the pallid men of Wall Street.

Suddenly in the distance he sees a human figure. Time was that his predecessors would have stopped to discuss the situation and its dangers, for the sight of one Indian suggested the presence of more, and the question came, were these friendly or fierce? But now the sled hurries on. It is only an Indian or half-breed hunter minding his traps, of which he may have a sufficient number to give him a circuit of ten or more miles away from and back to his lodge or village. He is approached and hailed by the driver, and with some pretty name very often— one that may mean in English "hawk flying across the sky when the sun is setting," or "blazing sun," or whatever. On goes the sled, and perhaps a village is the next object of interest; not a village in our sense of the word, but now and then a tepee or a hut peeping above the brush beside the water, the eye being led to them by the signs of slothful disorder close by —the rotting canoe frame, the bones, the dirty tattered blankets, the twig-formed skeleton of a steam bath, such as Indians resort to when tired or sick or uncommonly dirty, the worn-out snow-shoes hung on a tree, and the racks of frozen fish or dried meat here and there. A dog rushes down to the water-side barking furiously—an Indian dog of the currish type of paupers' dogs the world around—and this stirs the village pack, and brings out the squaws, who are addressed, as the trapper up the stream was, by some poetic names, albeit poetic license is sometimes strained to form names not at all pretty to polite senses, "All Stomach" being that of one dusky princess, and serving to indicate the lengths to which poesy may lead the untrammelled mind.

The sun sinks early, and if our traveller be journeying in the West and be a lover of nature, heaven send that his face be turned towards the sunset! Then, be the sky anything but completely storm-draped, he will see a sight so glorious that eloquence becomes a naked suppliant for alms beyond the gift of language when set to describe it. A few clouds are necessary to its perfection, and then they take on celestial dyes, and one sees, above the vanished sun, a blaze of golden yellow thinned into a tone that is luminous crystal. This is flanked by belts and breasts of salmon and ruby red, and all melt towards the zenith into a rose tone that has body at the base, but pales at top into a mere blush. This I have seen night after night on the lakes and the plains and on the mountains. But as the glory of it beckons the traveller ever towards itself, so the farther he follows, the more brilliant and gaudy will be his reward. Beyond the mountains the valleys and waters are more and more enriched, until, at the Pacific, even San Francisco's shabby sand-hills stir poetry and reverence in the soul by their borrowed magnificence.

The travellers soon stop to camp for the night, and while the "breed" falls to at the laborious but quick and simple work, the factor either helps or smokes his pipe. A sight-seer or sportsman would have set his man to bobbing for jack-fish or lake trout, or would have stopped a while to bag a partridge, or might have bought whatever of this sort the trapper or Indian village boasted, but, ten to one, this meal would be of bacon and bread or dried meat, and perhaps some flapjacks, such as would bring coin to a doctor in the city, but which seem ethereal and delicious in the wilderness, particularly if made half an inch thick, saturated with grease, well browned, and eaten while at the temperature and consistency of molten lava.

The sled is pulled up by the bank, the ground is cleared for a fire, wood and brush are cut, and the deft laborer starts the flame in a tent-like pyramid of kindlings no higher or broader than a teacup. This tiny fire he spreads by adding fuel until he has constructed and led up to a conflagration of logs as thick as his thighs, cleverly planned with a backlog and glowing fire bed, and a sapling bent over the hottest part to hold a pendent kettle on its tip. The dogs will have needed disciplining long before this, and if the driver be like many of his kind, and works himself into a fury, he will not hesitate to seize one and send his teeth together through its hide after he has beaten it until he is tired. The point of order having thus been raised and carried, the shaggy, often handsome, animals will be minded to forget their private grudges and quarrels, and, seated on their haunches, with their intelligent faces towards the fire, will watch the cooking intently. The pocket-knives or sheath-knives of the men will be apt to be the only table implement in use at the meal. Canada had reached the possession of seigniorial man-

sions of great character before any other knife was brought to table, though the ladies used costly blades set in precious and beautiful handles. To-day the axe ranks the knife in the wilderness, but he who has a knife can make and furnish his own table— and his house also, for that matter.

Supper over, and a glass of grog having been put down, with water from the hole in the ice whence the liquid for the inevitable tea was gotten, the night's rest is begun. The method for this varies. As good men as ever walked have asked nothing more cosey than a snug warm trough in the snow and a blanket or a robe; but perhaps this traveller will call for a shake-down of balsam boughs, with all the furs out of the sled for his covering. If nicer yet, he may order a low hollow chamber of three sides of banked snow, and a superstructure of crotched sticks and cross-poles, with canvas thrown over it. Every man to his quality, of course, and that of the servant calls for simply a blanket. With that he sleeps as soundly as if he were Santa Claus and only stirred once a year. Then will fall upon what seems the whole world the mighty hush of the wilderness, broken only occasionally by the hoot of an owl, the cry of a wolf, the deep thug of the straining ice on the lake, or the snoring of the men and dogs. But if the earth seems asleep, not so the sky. The magic shuttle of the aurora borealis is ofttimes at work up over that North country, sending its shifting lights weaving across the firmament with a tremulous brilliancy and energy we in this country get but pale hints of when we see the phenomenon at all. Flashing and palpitating incessantly, the rose-tinted waves and luminous white bars leap across the sky or dart up and down it in manner so fantastic and so forceful, even despite their shadowy thinness, that travellers have fancied themselves deaf to some seraphic sound that they believed such commotion must produce.

An incident of this typical journey I am describing would, at more than one season, be a meeting with some band of Indians going to a post with furs for barter. Though the bulk of these hunters fetch their quarry in the spring and early summer, some may come at any time. The procession may be only that of a family or of the two or more families that live together or as neighbors. The man, if there is but one group, is certain to be stalking ahead, carrying nothing but his gun. Then come the women, laden like pack-horses. They may have a sled packed with the furs and drawn by a dog or two, and an extra dog may bear a balanced load on his back, but the squaw is certain to have a spine-warping burden of meat and a battered kettle and a pappoose, and whatever personal property of any and every sort she and her liege lord own. Children who can walk have to do so, but it sometimes happens that a baby a year and a half or two years old is on her back, while a newborn infant, swaddled in blanket stuff, and bagged and tied like a Bologna sausage, surmounts the load on the sled. A more tatterdemalion outfit than a band of these pauperized savages form it would be difficult to imagine. On the plains they will have horses dragging travoises, dogs with travoises, women and children loaded with impedimenta, a colt or two running loose, the lordly men riding free, straggling curs a plenty, babies in arms, babies swaddled, and toddlers afoot, and the whole battalion presenting at its exposed points exhibits of torn blankets, raw meat, distorted pots and pans, tent, poles, and rusty traps, in all eloquently suggestive of an eviction in the slums of a great city.

I speak thus of these people not willingly, but out of the necessity of truth-telling. The Indian east of the Rocky Mountains is to me the subject of an admiration which is the stronger the more nearly I find him as he was in his prime. It is not his fault that most of his race have degenerated. It is not our fault that we have better uses for the continent than those to which he put it. But it is our fault that he is, as I have seen him, shivering in a cotton tepee full of holes, and turning around and around before a fire of wet wood to keep from freezing to death; furnished meat if he has been fierce enough to make us fear him, left to starve if he has been docile; taught, aye, forced to beg, mocked at by a religion he cannot understand, from the mouths of men who apparently will not understand him; debauched with rum, despoiled by the lust of white men in every form that lust can take. Ah, it is a sickening story. Not in Canada, do you say? Why, in the northern wilds of Canada are districts peopled by beggars who have been in such pitiful stress for food and covering that the Hudson Bay Company has kept them alive with advances of provisions and blankets winter after winter. They are Indians who in their strength never gave the Government the concern it now fails to show for their weakness. The great fur company has thus added generosity to its long career of just dealing with these poor adult children; for it is a fact that though the company has made what profit it might, it has not, in a century at least, cheated the Indians, or made false representations to them, or lost their good-will and respect by any feature of its policy towards them. Its relation to them has been paternal, and they owe none of their degradation to it.

I have spoken of the visits of the natives to the posts. There are two other arrivals of great consequence—the coming of the supplies, and of the winter mail or packet. I have seen the provisions and trade goods being put up in bales in the great mercantile storehouse of the company in Winnipeg —a store like a combination of a Sixth Avenue ladies' bazaar and one of our wholesale grocers' shops —and I have seen such weights of canned vegetables and canned plum-pudding and bottled ale and other luxuries that I am sure that in some posts there is good living on high days and holidays if not always. The stores are packed in parcels averaging sixty pounds (and sometimes one hundred), to make them convenient for handling on the portages—"for packing them over the carries," as our traders used to say. It is in following these supplies that we become most keenly sensible of the changes time has wrought in the methods of the company. The day was, away back in the era of the Northwest Company, that the goods for the posts went up the Ottawa

HALT OF A YORK BOAT BRIGADE FOR THE NIGHT

from Montreal in great canoes manned by hardy *voyageurs* in picturesque costumes, wielding scarlet paddles, and stirring the forests with their happy songs. The scene shifted, the companies blended, and the centre of the trade moved from old Fort William, close to where Port Arthur now is on Lake Superior, up to Winnipeg, on the Red River of the North. Then the Canadians and their cousins, the half-breeds, more picturesque than ever, and manning the great York boats of the Hudson Bay Company, swept in a long train through Lake Winnipeg to Norway House, and thence by a marvellous water route all the way to the Rockies and the Arctic, sending off freight for side districts at fixed points along the course. The main factories on this line, maintained as such for more than a century, bear names whose very mention stirs the blood of one who knows the romantic, picturesque, and poetic history and atmosphere of the old company when it was the landlord (in part, and in part monopolist) of a territory that cut into our Northwest and Alaska, and swept from Labrabor to Vancouver Island. Northward and westward, by waters emptying into Hudson Bay, the brigade of great boats worked through a region embroidered with sheets and ways of water. The system that was next entered, and which bore more nearly due west, bends and bulges with lakes and straits like a ribbon all curved and knotted. Thus, at a great portage, the divide was reached and crossed; and so the waters flowing to the Arctic, and one — the Peace River — rising beyond the Rockies, were met and travelled. This was the way and the method until after the Canadian Pacific Railway was built, but now the Winnipeg route is of subordinate importance, and feeds only the region near the west side of Hudson Bay. The Northern supplies now go by rail from Calgary, in Alberta, over the plains by the new Edmonton railroad. From Edmonton the goods go by cart to Athabasca Landing, there to be laden on a steamboat, which takes them northward until some rapids are met, and avoided by the use of a singular combination of bateaux and tramway rails. After a slow progress of fifteen miles another steamboat is met, and thence they follow the Athabasca, through Athabasca Lake, and so on up to a second rapids, on the Great Slave River this time, where oxen and carts carry them across a sixteen-mile portage to a screw steamer, which finishes the 3000-mile journey to the North. Of course the shorter branch routes, distributing the goods on either side of the main track, are still traversed by canoes and hardy fellows in the old way, but with shabby accessories of costume and spirit. These boatmen, when they come to a portage, produce their tomplines, and "pack" the goods to the next waterway. By means of these "lines" they carry great weights, resting on their backs, but supported from their skulls, over which the strong straps are passed.

The winter mail-packet, starting from Winnipeg in the depth of the season, goes to all the posts by dog train. The letters and papers are packed in great boxes and strapped to the sleds, beside or behind which the drivers trot along, cracking their lashes and pelting and cursing the dogs. A more direct course than the old Lake Winnipeg way has usually been followed by this packet; but it is thought that the route *via* Edmonton and Athabasca Landing will serve better yet, so that another change may be made. This is a small exhibition as compared with the brigade that takes the supplies, or those others that come plashing down the streams and across the country with the furs every year. But only fancy how eagerly this solitary semi-annual mail is waited for! It is a little speck on the snow-wrapped upper end of all North America. It cuts a tiny trail, and here and there lesser black dots move off from it to cut still slenderer threads, zigzagging to the side factories and lesser posts; but we may be sure that if human eyes could see so far, all those of the white men in all that vast tangled system of trading centres would be watching the little caravan, until at last each pair fell upon the expected missives from the throbbing world this side of the border.

VIII

CANADA'S EL DORADO

THERE is on this continent a territory of imperial extent which is one of the Canadian sisterhood of States, and yet of which small account has been taken by those who discuss either the most advantageous relations of trade or that closer intimacy so often referred to as a possibility in the future of our country and its northern neighbor. Although British Columbia is advancing in rank among the provinces of the Dominion by reason of its abundant natural resources, it is not remarkable that we read and hear little concerning it. The people in it are few, and the knowledge of it is even less in proportion. It is but partially explored, and for what can be learned of it one must catch up information piecemeal from blue-books, the pamphlets of scientists, from tales of adventure, and from the less trustworthy literature composed to attract travellers and settlers.

It would severely strain the slender facts to make a sizable pamphlet of the history of British Columbia. A wandering and imaginative Greek called Juan de Fuca told his people that he had discovered a passage from ocean to ocean between this continent and a great island in the Pacific. Sent there to seize and fortify it, he disappeared—at least from history. This was about 1592. In 1778 Captain Cook roughly surveyed the coast, and in 1792 Captain Vancouver, who as a boy had been with Cook on two voyages, examined the sound between the island and the mainland with great care, hoping to find that it led to the main water system of the interior. He gave to the strait at the entrance the nickname of the Greek, and in the following year received the transfer of authority over the country from the Spanish commissioner Bodega of Quadra, then established there. The two put aside false modesty, and named the great island "the Island of Vancouver and Quadra." At the time the English sailor was there it chanced that he met that hardy old homespun baronet Sir Alexander Mackenzie, who was the first man to cross the continent, making the astonishing journey in a canoe manned by Iroquois Indians. The main-land became known as New Caledonia. It took its present name from the Columbia River, and that, in turn, got its name from the ship *Columbia*, of Boston, Captain Gray, which entered its mouth in 1792, long after the Spaniards had known the stream and called it the Oregon. The rest is quickly told. The region passed into the hands of the fur-traders. Vancouver Island became a crown colony in 1849, and

British Columbia followed in 1858. They were united in 1866, and joined the Canadian confederation in 1871. Three years later the province exceeded both Manitoba and Prince Edward Island in the value of its exports, and also showed an excess of exports over imports. It has a Lieutenant-governor and Legislative Assembly, and is represented at Ottawa in accordance with the Canadian system. Its people have been more closely related to ours in business than those of any other province, and they entertain a warm friendly feeling towards "the States." In the larger cities the Fourth of July is informally but generally observed as a holiday.

British Columbia is of immense size. It is as extensive as the combination of New England, the Middle States and Maryland, the Virginias, the Carolinas, and Georgia, leaving Delaware out. It is larger than Texas, Colorado, Massachusetts, and New Hampshire joined together. Yet it has been all but overlooked by man, and may be said to be an empire with only one wagon road, and that is but a blind artery halting in the middle of the country. But whoever follows this necessarily incomplete survey of what man has found that region to be, and of what his yet puny hands have drawn from it, will dismiss the popular and natural suspicion that it is a wilderness worthy of its present fate. Until the whole globe is banded with steel rails and yields to the plough, we will continue to regard whatever region lies beyond our doors as waste-land, and to fancy that every line of latitude has its own unvarying climatic characteristics. There is an opulent civilization in what we once were taught was "the Great American Desert," and far up at Edmonton, on the Peace River, farming flourishes despite the fact that it is where our school-books located a zone of perpetual snow. Farther along we shall study a country crossed by the same parallels of latitude that dissect inhospitable Labrador, and we shall discover that as great a difference exists between the two shores of the continent on that zone as that which distinguishes California from Massachusetts. Upon the coast of this neglected corner of the world we shall see that a climate like that of England is produced, as England's is, by a warm current in the sea; in the southern half of the interior we shall discover valleys as inviting as those in our New England; and far north, at Port Simpson, just below the down reaching claw of our Alaska, we shall find such a climate as Halifax enjoys.

British Columbia has a length of 800 miles, and averages 400 miles in width. To whoever crosses the country it seems the scene of a vast earth-disturbance, over which mountains are scattered without system. In fact, however, the Cordillera belt is there divided into four ranges, the Rockies forming the eastern boundary, then the Gold Range, then the Coast Range, and, last of all, that partially submerged chain whose upraised parts form Vancouver and the other mountainous islands near the main-land in the Pacific. A vast valley flanks the south-western side of the Rocky Mountains, accompanying them from where they leave our North-western States in a wide straight furrow for a distance of 700 miles. Such great rivers as the Columbia, the Fraser, the Parsnip, the Kootenay, and the Finlay are encountered in it. While it has a lesser agricultural value than other valleys in the province, its mineral possibilities are considered to be very great, and when, as must be the case, it is made the route of communication between one end of the territory and the other, a vast timber supply will be rendered marketable.

The Gold Range, next to the westward, is not bald, like the Rockies, but, excepting the higher peaks, is timbered with a dense forest growth. Those busiest of all British Columbian explorers, the "prospectors," have found much of this system too difficult even for their pertinacity. But the character of the region is well understood. Here are high plateaus of rolling country, and in the mountains are glaciers and snow fields. Between this system and the Coast Range is what is called the Interior Plateau, averaging one hundred miles in width, and following the trend of that portion of the continent, with an elevation that grows less as the north is approached. This plateau is crossed and followed by valleys that take every direction, and these are the seats of rivers and watercourses. In the southern part of this plateau is the best grazing land in the province, and much fine agricultural country, while in the north, where the climate is more moist, the timber increases, and parts of the land are thought to be convertible into farms. Next comes the Coast Range, whose western slopes are enriched by the milder climate of the coast; and beyond lies the remarkably tattered shore of the Pacific, lapped by a sheltered sea, verdant, indented by numberless inlets, which, in turn, are faced by uncounted islands, and receive the discharge of almost as many streams and rivers—a wondrously beautiful region, forested by giant trees, and resorted to by numbers of fish exceeding calculation and belief. Beyond the coast is the bold chain of mountains of which Vancouver Island and the Queen Charlotte Islands are parts. Here is a vast treasure in that coal which our naval experts have found to be the best on the Pacific coast, and here also are traces of metals, whose value industry has not yet established.

It is a question whether this vast territory has yet 100,000 white inhabitants. Of Indians it has but 20,000, and of Chinese about 8000. It is a vast land of silence, a huge tract slowly changing from the field and pleasure-ground of the fur-trader and sportsman to the quarry of the miner. The Canadian Pacific Railway crosses it, revealing to the immigrant and the globe-trotter an unceasing panorama of grand, wild, and beautiful scenery unequalled on this continent. During a few hours the traveller sees, across the majestic cañon of the Fraser, the neglected remains of the old Cariboo stage road, built under pressure of the gold craze. It demonstrated surprising energy in the baby colony, for it connected Yale, at the head of short steam navigation on the Fraser, with Barkerville, in the distant Cariboo country, 400 miles away, and it cost $500,000. The traveller sees here and there an Indian village or a "mission," and now and then a tiny town; but for the most part his eye scans only the primeval forest, lofty mountains, valleys covered with trees as beasts are with fur, cascades, turbulent streams, and huge sheltered lakes. Except at the stations, he sees few men. Now he notes a group of Chinamen at work on the railway; anon he sees an Indian upon a clumsy perch and searching the Fraser for salmon, or in a canoe paddling towards the gorgeous sunset that confronts the daily west-bound train as it rolls by great Shuswap Lake.

But were the same traveller out of the train, and gifted with the power to make himself ubiquitous, he would still be, for the most part, lonely. Down in the smiling bunch-grass valleys in the south he would see here and there the outfit of a farmer or the herds of a cattle-man. A burst of noise would astonish him near by, in the Kootenay country, where the new silver mines are being worked, where claims have been taken up by the thousand, and whither a railroad is hastening. Here and there, at points out of sight one from another, he would hear the crash of a lumberman's axe, the report of a hunter's rifle, or the crackle of an Indian's fire. On the Fraser he would find a little town called Yale, and on the coast the streets and ambitious buildings and busy wharves of Vancouver would astonish him. Victoria, across the strait, a town of larger size and remarkable beauty, would give him company, and near Vancouver and Victoria the little cities of New Westminster and Nanaimo (lumber and coal ports respectively) would rise before him. There, close together, he would see more than half the population of the province.

AN IMPRESSION OF SHUSWAP LAKE, BRITISH COLUMBIA

Fancy his isolation as he looked around him in the northern half of the territory, where a few trails lead to fewer posts of the Hudson Bay Company, where the endless forests and multitudinous lakes and streams are cut by but infrequent paddles in the hands of a race that has lost one-third its numerical strength in the last ten years, where the only true homes are within the palisades or the unguarded log-cabin of the fur-trading agents, and where the only other white men are either washing sand in the river bars, driving the stages of the only line that penetrates a piece of the country, or are those queer devil-may-care but companionable Davy Crocketts of the day who are guides now and then, hunters half the time, placer-miners when they please, and whatever else there is a call for betweentimes!

A very strange sight that my supposititious traveller would pause long to look at would be the herds of wild horses that defy the Queen, her laws, and her subjects in the Lillooet Valley. There are thousands of them there, and over in the Nicola and Chilcotin country, on either side of the Fraser, north of Washington State. They were originally of good stock, but now they not only defy capture, but eat valuable grass, and spoil every horse turned out to graze. The newspapers aver that the Government must soon be called upon to devise means for ridding the valleys of this nuisance. This is one of those sections which promise well for future stock-raising and agricultural operations. There are plenty such. The Nicola Valley has been settled twenty years, and there are many cattle there, on numerous ranches. It is good land, but rather high for grain, and needs irrigation. The snowfall varies greatly in all these valleys, but in ordinary winters horses and cattle manage well with four to six weeks' feeding. On the upper Kootenay, a valley eight to ten miles wide, ranching began a quarter of a century ago, during the gold excitement. The "cow-men" raise grain for themselves there. This valley is 3000 feet high. The Okanagon Valley is lower, and is only from two to five miles wide, but both are of similar character, of very great length, and are crossed and intersected by branch valleys. The greater part of the Okanagon does not need irrigating. A beautiful country is the Kettle River region, along the boundary between the Columbia and the Okanagon. It is narrow, but flat and smooth on the bottom, and the land is very fine. Bunch-grass covers the hills around it for a distance of from four hundred to five hundred feet, and there timber begins. It is only in occasional years that the Kettle River Valley needs water. In the Spallumcheen Valley one farmer had 500 acres in grain last summer, and the most modern agricultural machinery is in use there. These are mere notes of a few among almost innumerable valleys that are clothed with bunch-grass, and that often possess the characteristics of beautiful parks. In many wheat can be and is raised, possibly in most of them. I have notes of the successful growth of peaches, and of the growth of almond-trees to a height of fourteen feet in four years, both in the Okanagon country.

The shooting in these valleys is most alluring to those who are fond of the sport. Caribou, deer, bear, prairie-chicken, and partridges abound in them. In all probability there is no similar extent of country that equals the valley of the Columbia, from which, in the winter of 1888, between six and eight tons of deer-skins were shipped by local traders, the result of legitimate hunting. But the forests and mountains are as they were when the white man first saw them, and though the beaver and sea-otter, the marten, and those foxes whose furs are coveted by the rich, are not as abundant as they once were, the rest of the game is most plentiful. On the Rockies and on the Coast Range the mountain-goat, most difficult of beasts to hunt, and still harder to get, is abundant yet. The "big-horn," or mountain-sheep, is not so common, but the hunting thereof is usually successful if good guides are obtained. The cougar, the grizzly, and the lynx are all plentiful, and black and brown bears are very numerous. Elk are going the way of the "big-horn"—are preceding that creature, in fact. Pheasants (imported), grouse, quail, and water-fowl are among the feathered game, and the river and lake fishing is such as is not approached in any other part of the Dominion. The province is a sportsman's Eden, but the hunting of big game there is not a venture to be lightly undertaken. It is not alone the distance or the cost that gives one pause, for, after the province is reached, the mountain-climbing is a task that no amount of wealth will lighten. And these are genuine mountains, by-the-way, wearing eternal caps of snow, and equally eternal deceit as to their distances, their heights, and as to all else concerning which a rarefied atmosphere can hocus-pocus a stranger. There is one animal, king of all the beasts, which the most unaspiring hunter may chance upon as well as the bravest, and that animal carries a perpetual chip upon its shoulder, and seldom turns from an encounter. It is the grizzly-bear. It is his presence that gives you either zest or pause, as you may decide, in hunting all the others that roam the mountains. Yet, in that hunter's dream-land it is the grizzly that attracts many sportsmen every year.

From the headquarters of the Hudson Bay Company in Victoria I obtained the list of animals in whose skins that company trades at that station. It makes a formidable catalogue of zoological products, and is as follows: Bears (brown, black, grizzly), beaver, badger, foxes (silver, cross, and red), fishers, martens, minks, lynxes, musk-rat, otter (sea or land), panther, raccoon, wolves (black, gray, and coyote), black-tailed deer, stags (a true stag, growing to the size of an ox, and found on the hills of Vancouver Island), caribou or reindeer, hares, mountain-goat, big-horn (or mountain-sheep), moose (near the Rockies), wood-buffalo (found in the north, not greatly different from the bison, but larger), geese, swans, and duck.

The British Columbian Indians are of such unprepossessing appearance that one hears with comparative equanimity of their numbering only 20,000 in all, and of their rapid shrinkage, owing principally to the vices of their women. They are, for the most part, canoe Indians, in the interior as well as on the coast, and they are (as one might suppose a nation of tailors would become) short-legged, and with those limbs small and inclined to bow. On the other hand, their exercise with the paddle has given them a disproportionate development of their shoulders and chests, so that, being too large above and too small below, their appearance is very peculiar. They are fish-eaters the year around; and though some, like the Hydahs upon the coast, have been warlike and turbulent, such is not the reputation of those in the interior. It was the meat-eating Indian who made war a vocation and self-torture a dissipation. The fish-eating Indian kept out of his way. These short squat British Columbian natives are very dark-skinned, and have physiognomies so different from those of the Indians east of the Rockies that the study of their faces has tempted the ethnologists into extraordinary guessing upon their origin, and into a contention which I prefer to avoid. It is not guessing to say that their high cheek-bones and flat faces make them resemble the Chinese. That is true to such a degree that in walking the streets of Victoria, and meeting alternate Chinamen and Siwash, it is not always easy to say which is which, unless one proceeds upon the assumption that if a man looks clean he is apt to be a Chinaman, whereas if he is dirty and ragged he is most likely to be a Siwash.

You will find that seven in ten among the more intelligent British Columbians conclude these Indians to be of Japanese origin. The Japanese current is neighborly to the province, and it has drifted Japanese junks to these shores. When the first traders visited the neighborhood of the mouth of the Columbia they found beeswax in the sand near the vestiges of a wreck, and it is said that one wreck of a junk was met with, and 12,000 pounds of this wax was found on her. Whalers are said to have frequently encountered wrecked and drifting junks in the eastern Pacific, and a local legend has it that in 1834 remnants of a junk with three Japanese and a cargo of pottery were found on the coast south of Cape Flattery. Nothing less than all this should excuse even a rudderless ethnologist for so cruel a reflection upon the Japanese, for these Indians are so far from pretty that all who see them agree with Captain Butler, the traveller, who wrote that "if they are of the Mongolian type, the sooner the Mongolians change their type the better."

The coast Indians are splendid sailors, and their dugouts do not always come off second best in racing with the boats of white men. With a primitive yet ingeniously made tool, like an adze, in the construction of which a blade is tied fast to a bent handle of bone, these natives laboriously pick out the heart of a great cedar log, and shape its outer sides into the form of a boat. When the log is properly hollowed, they fill it with water, and then drop in stones which they have heated in a fire. Thus they steam the boat so that they may spread the sides and fit in the crossbars which keep it strong and preserve its shape. These dugouts are sometimes sixty feet long, and are used for whaling and long voyages in rough seas. They are capable of carrying tons of the salmon or oolachan or herring, of which these people, who live as their fathers did, catch sufficient in a few days for their maintenance throughout a whole year. One gets an idea of the swarms of fish that infest those waters by the knowledge that before nets were used the herring and the oolachan, or candle-fish, were swept into these boats by an implement formed by studding a ten-foot pole with spikes or nails. This was swept among the fish in the water, and the boats were speedily filled with the creatures that were impaled upon the spikes. Salmon, sea-otter, otter, beaver, marten, bear, and deer (or caribou or moose) were and still are the chief resources of

THE TSCHUMMUM, OR TOOL USED IN MAKING CANOES

most of the Indians. Once they sold the fish and
the peltry to the Hudson Bay Company, and ate
what parts or surplus they did not sell. Now they
work in the canneries or fish for them in summer,
and hunt, trap, or loaf the rest of the time. However,
while they still fish and sell furs, and while some are
yet as their fathers were, nearly all the coast Indians
are semi-civilized. They have at least the white
man's clothes and hymns and vices. They have
churches; they live in houses; they work in canneries. What little there was that was picturesque
about them has vanished only a few degrees faster
than their own extinction as a pure race, and they
are now a lot of longshoremen. What Mr. Duncan
did for them in Metlakahtla—especially in housing
the families separately—has not been arrived at even
in the reservation at Victoria, where one may still
see one of the huge, low, shed-like houses they prefer,
ornamented with totem poles, and arranged for eight
families, and consequently for a laxity of morals for
which no one can hold the white man responsible.

They are a tractable people, and take as kindly
to the rudiments of civilization, to work, and to co-
operation with the whites as the plains Indian does
to tea, tobacco, and whiskey. They are physically
but not mentally inferior to the plainsman. They
carve bowls and spoons of stone and bone, and their
heraldic totem poles are cleverly shapen, however
grotesque they may be. They still make them, but
they oftener carve little ones for white people, just as
they make more silver bracelets for sale than for wear.
They are clever at weaving rushes and cedar bark

into mats, baskets, floor-cloths, and cargo covers. In a word, they were more prone to work at the outset than most Indians, so that the present longshore career of most of them is not greatly to be wondered at.

To any one who threads the vast silent forests of the interior, or journeys upon the trafficless waterways, or, gun in hand, explores the mountains for game, the infrequency with which Indians are met becomes impressive. The province seems almost unpeopled. The reason is that the majority of the Indians were ever on the coast, where the water yielded food at all times and in plenty. The natives of the interior were not well fed or prosperous when the first white men found them, and since then smallpox, measles, vice, and starvation have thinned them terribly. Their graveyards are a feature of the scenery which all travellers in the province remember. From the railroad they may be seen along the Fraser, each grave apparently having a shed built over it, and a cross rising from the earth beneath the shed. They had various burial customs, but a majority buried their dead in this way, with queerly-carved or painted sticks above them, where the cross now testifies to the work at the "missions." Some Indians marked a man's burial-place with his canoe and his gun; some still box their dead and leave the boxes on top of the earth, while others bury the boxes. Among the southern tribes a man's horse was often killed, and its skin decked the man's grave; while in the far north it was the custom among the Stickeens to slaughter the personal attendants of a chief when he died. The Indians along the Skeena River cremated their dead, and sometimes hung the ashes in boxes to the family totem pole. The Hydahs, the fierce natives of certain of the islands, have given up cremation, but they used to believe that if they did not burn a man's body their enemies would make charms from it. Polygamy flourished on the coast, and monogamy in the interior, but the contrast was due to the difference in the worldly wealth of the Indians. Wives had to be bought and fed, and the woodsmen could only afford one apiece.

To return to their canoes, which most distinguish them. When a dugout is hollowed and steamed, a prow and stern are added of separate wood. The prow is always a work of art, and greatly beautifies the boat. It is in form like the breast, neck, and bill of a bird, but the head is intended to represent that of a savage animal, and is so painted. A mouth is cut into it, ears are carved on it, and eyes are painted on the sides; bands of gay paint are put upon the neck, and the whole exterior of the boat is then painted red or black, with an ornamental line of another color along the edge or gunwale. The sailors sit upon the bottom of the boat, and propel it with paddles. Upon the water these swift vessels, with their fierce heads uplifted before their long, slender bodies, appear like great serpents or nondescript marine monsters, yet they are pretty and graceful withal. While still holding aloof from the ethnologists' contention, I yet may add that a bookseller in Victoria came into the possession of a packet of photographs taken by an amateur traveller in the interior of China, and on my first visit to the province, nearly four years ago, I found, in looking through these views, several Chinese boats which were strangely and remarkably like the dugouts of the provincial Indians. They were too small in the pictures for it to be possible to decide whether they were built up or dug out, but in general they were of the same external appearance, and each one bore the upraised animal-head prow, shaped and painted like those I could see one block away from the bookseller's shop in Victoria. But such are not the canoes used by the Indians of the interior. From the Kootenay near our border to the Cassiar in the far north, a cigar-shaped canoe seems to be the general native vehicle. These are sometimes made of a

THE FIRST OF THE SALMON RUN, FRASER RIVER

sort of scroll of bark, and sometimes they are dugouts made of cotton-wood logs. They are narrower than either the cedar dugouts of the coast or the birch-bark canoes of our Indians, but they are roomy, and fit for the most dangerous and deft work in threading the rapids which everywhere cut up the navigation of the streams of the province into separated reaches. The Rev. Dr. Gordon, in his notes upon a journey in this province, likens these canoes to horse-troughs, but those I saw in the Kootenay country were of the shape of those cigars that are pointed at both ends.

Whether these canoes are like any in Tartary or China or Japan, I do not know. My only quest for special information of that character proved disappointing. One man in a city of British Columbia is said to have studied such matters more deeply and to more purpose than all the others, but those who referred me to him cautioned me that he was eccentric.

"You don't know where these Indians came from, eh?" the *savant* replied to my first question. "Do you know how oyster-shells got on top of the Rocky Mountains? You don't, eh? Well, I know a woman who went to a dentist's yesterday to have eighteen teeth pulled. Do you know why women prefer artificial teeth to those which God has given them? You don't, eh? Why, man, you don't know anything."

While we were—or he was—conversing, a laboring-man who carried a sickle came to the open door, and was asked what he wanted.

"I wish to cut your thistles, sir," said he.

"Thistles?" said the *savant*, disturbed at the interruption. "—— the thistles! We are talking about Indians."

Nevertheless, when the laborer had gone, he had left the subject of thistles uppermost in the *savant's* mind, and the conversation took so erratic a turn that it might well have been introduced hap-hazard into *Tristram Shandy*.

"About thistles," said the *savant*, laying a gentle hand upon my knee. "Do you know that they are the Scotchmen's totems? Many years ago a Scotchman, sundered from his native land, must needs set up his totem, a thistle, here in this country; and now, sir, the thistle is such a curse that I am haled up twice a year and fined for having them in my yard."

But nearly enough has been here said of the native population. Though the Indians boast dozens of tribal names, and almost every island on the coast and village in the interior seems the home of a separate tribe, they will be found much alike — dirty, greasy, sore-eyed, short-legged, and with their unkempt hair cut squarely off, as if a pot had been upturned over it to guide the operation. The British Columbians do not bother about their tribal divisions, but use the old traders' Chinook terms, and call every male a "siwash" and every woman a "klootchman."

Since the highest Canadian authority upon the subject predicts that the northern half of the Cordilleran ranges will admit of as high a metalliferous development as that of the southern half in our Pacific

States, it is important to review what has been done in mining, and what is thought of the future of that industry in the province. It may almost be said that the history of gold-mining there is the history of British Columbia. Victoria, the capital, was a Hudson Bay post established in 1843, and Vancouver, Queen Charlotte's, and the other islands, as well as the mainland, were of interest to only a few white men as parts of a great fur-trading field with a small Indian population. The first nugget of gold was found at what is now called Gold Harbor, on the west coast of the Queen Charlotte Islands, by an Indian woman, in 1851. A part of it, weighing four or five ounces, was taken by the Indians to Fort Simpson and sold. The Hudson Bay Company, which has done a little in every line of business in its day, sent a brigantine to the spot, and found a quartz vein traceable eighty feet, and yielding a high percentage of gold. Blasting was begun, and the vessel was loaded with ore; but she was lost on the return voyage. An American vessel, ashore at Esquimault, near Victoria, was purchased, renamed the *Recovery*, and sent to Gold Harbor with thirty miners, who worked the vein until the vessel was loaded and sent to England. News of the mine travelled, and in another year a small fleet of vessels came up from San Francisco; but the supply was seen to be very limited, and after $20,000 in all had been taken out, the field was abandoned.

In 1855 gold was found by a Hudson Bay Company's employé at Fort Colville, now in Washington State, near the boundary. Some Thompson River

(B. C.) Indians who went to Walla Walla spread a report there that gold, like that discovered at Colville, was to be found in the valley of the Thompson. A party of Canadians and half-breeds went to the region referred to, and found placers nine miles above the mouth of the river. By 1858 the news and the authentication of it stirred the miners of California, and an astonishing invasion of the virgin province began. It is said that in the spring of 1858 more than twenty thousand persons reached Victoria from San Francisco by sea, distending the little fur-trading post of a few hundred inhabitants into what would even now be called a considerable city; a city of canvas, however. Simultaneously a third as many miners made their way to the new province on land. But the land was covered with mountains and dense forests, the only route to its interior for them was the violent, almost boiling, Fraser River, and there was nothing on which the lives of this horde of men could be sustained. By the end of the year out of nearly thirty thousand adventurers only a tenth part remained. Those who did stay worked the river bars of the lower Fraser until in five months they had shipped from Victoria more than half a million dollars' worth of gold. From a historical point of view it is a peculiar coincidence that in 1859, when the attention of the world was thus first attracted to this new country, the charter of the Hudson Bay Company expired, and the territory passed from its control to become like any other crown colony.

In 1860 the gold-miners, seeking the source of the "flour" gold they found in such abundance in the bed

of the river, pursued their search into the heart and
almost the centre of that forbidding and unbroken territory. The Quesnel River became the seat of their
operations. Two years later came another extraordinary immigration. This was not surprising, for 1500
miners had in one year (1861) taken out $2,000,000

INDIAN SALMON-FISHING IN THE THRASHER

in gold-dust from certain creeks in what is called the Cariboo District, and one can imagine (if one does not remember) what fabulous tales were based upon this fact. The second stampede was of persons from all over the world, but chiefly from England, Canada, Australia, and New Zealand. After that there were new "finds" almost every year, and the miners worked gradually northward until, about 1874, they had travelled through the province, in at one end and out at the other, and were working the tributaries of the Yukon River in the north, beyond the 60th parallel. Mr. Dawson estimates that the total yield of gold between 1858 and 1888 was $54,108,804; the average number of miners employed each year was 2775, and the average earnings per man per year were $622.

In his report, published by order of Parliament, Mr. Dawson says that while gold is so generally distributed over the province that scarcely a stream of any importance fails to show at least "colors" of the metal, the principal discoveries clearly indicate that the most important mining districts are in the systems of mountains and high plateaus lying to the south-west of the Rocky Mountains and parallel in direction with them.

This mountain system next to and south-west of the Rockies is called, for convenience, the Gold Range, but it comprises a complex belt "of several more or less distinct and partly overlapping ranges"—the Purcell, Selkirk, and Columbia ranges in the south, and in the north the Cariboo, Omenica, and Cassiar ranges. "This series or system constitutes

the most important metalliferous belt of the province.
The richest gold fields are closely related to it, and
discoveries of metalliferous lodes are reported in
abundance from all parts of it which have been explored. The deposits already made known are very
varied in character, including highly argentiferous
galenas and other silver ores and auriferous quartz
veins." This same authority asserts that the Gold
Range is continued by the Cabinet, Cœur d'Alene,
and Bitter Root mountains in our country. While
there is no single well-developed gold field as in California, the extent of territory of a character to occasion a hopeful search for gold is greater in the province than in California. The average man of business
to whom visitors speak of the mining prospects of the
province is apt to declare that all that has been lacking is the discovery of one grand mine and the enlistment of capital (from the United States, they generally say) to work it. Mr. Dawson speaks to the
same point, and incidentally accounts for the retarded
development in his statement that one noteworthy
difference between practically the entire area of the
province and that of the Pacific States has been occasioned by the spread and movement of ice over the
province during the glacial period. This produced
changes in the distribution of surface materials and
directions of drainage, concealed beneath "drifts" the
indications to which prospectors farther south are used
to trust, and by other means obscured the outcrops of
veins which would otherwise be well marked. The
dense woods, the broken navigation of the rivers, in
detached reaches, the distance from the coast of the

GOING TO THE POTLATCH—BIG CANOE, NORTH-WEST COAST

richest districts, and the cost of labor supplies and machinery—all these are additional and weighty reasons for the slowness of development. But this was true of the past and is not of the present, at least so far as southern British Columbia is concerned. Railroads are reaching up into it from our country and down from the transcontinental Canadian Railway, and capital, both Canadian and American, is rapidly swelling an already heavy investment in many new and promising mines. Here it is silver-mining that is achieving importance.

Other ores are found in the province. The iron which has been located or worked is principally on the islands—Queen Charlotte, Vancouver, Texada, and the Walker group. Most of the ores are magnetites, and that which alone has been worked—on Texada Island—is of excellent quality. The output of copper from the province is likely soon to become considerable. Masses of it have been found from time to time in various parts of the province—in the Vancouver series of islands, on the main-land coast, and in the interior. Its constant and rich association with silver shows lead to be abundant in the country, but it needs the development of transport facilities to give it value. Platinum is more likely to attain importance as a product in this than in any other part of North America. On the coast the granites are of such quality and occur in such abundance as to lead to the belief that their quarrying will one day be an important source of income, and there are marbles, sandstones, and ornamental stones of which the same may be said.

One of the most valuable products of the province is coal, the essential in which our Pacific coast States are the poorest. · The white man's attention was first attracted to this coal in 1835 by some Indians who brought lumps of it from Vancouver Island to the Hudson Bay post on the main-land, at Milbank Sound. The *Beaver*, the first steamship that stirred the waters of the Pacific, reached the province in 1836, and used coal that was found in outcroppings on the island beach. Thirteen years later the great trading company brought out a Scotch coal-miner to look into the character and extent of the coal find, and he was followed by other miners and the necessary apparatus for prosecuting the inquiry. In the mean time the present chief source of supply at Nanaimo, seventy miles from Victoria and about opposite Vancouver, was discovered, and in 1852 mining was begun in earnest. From the very outset the chief market for the coal was found to be San Francisco.

The original mines are now owned by the Vancouver Coal-mining and Land Company. Near them are the Wellington Mines, which began to be worked in 1871. Both have continued in active operation from their foundation, and with a constantly and rapidly growing output. A third source of supply has very recently been established with local and American capital in what is called the Comox District, back of Baynes Sound, farther north than Nanaimo, on the eastern side of Vancouver Island. These new works are called the Union Mines, and, if the predictions of my informants prove true, will produce an output equal to that of the older Nanaimo collieries combined. In 1884 the coal shipped from Nanaimo amounted to 1000 tons for every day of the year, and in 1889 the total shipment had reached 500,000 tons. As to the character of the coal, I quote again from Mr. Dawson's report on the minerals of British Columbia, published by the Dominion Government:

" Rocks of cretaceous age are developed over a considerable area in British Columbia, often in very great thickness, and fuels occur in them in important quantity in at least two distinct stages, of which the lower and older includes the coal measures of the Queen Charlotte Islands and those of Quatsino Sound on Vancouver Island, with those of Crow Nest Pass in the Rocky Mountains; the upper, the coal measures of Nanaimo and Comox, and probably also those of Suquash and other localities. The lower rocks hold both anthracite and bituminous coal in the Queen Charlotte Islands, but elsewhere contain bituminous coal only. The upper have so far been found to yield bituminous coal only. The fuels of the tertiary rocks are, generally speaking, lignites, but include also various fuels intermediate between these and true coals, which in a few places become true bituminous coals."

It is thought to be more than likely that the Comox District may prove far more productive than the Nanaimo region. It is estimated that productive measures underlie at least 300 square miles in the Comox District, exclusive of what may extend beyond the shore. The Nanaimo area is estimated at 200 square miles, and the product is no better than, if it equals, that of the Comox District.

Specimens of good coal have been found on the main-land in the region of the upper Skeena River, on the British Columbia water-shed of the Rockies near Crow Nest Pass, and in the country adjacent to the Peace River in the eastern part of the province.

Anthracite which compares favorably with that of Pennsylvania has been found at Cowgitz, Queen Charlotte Islands. In 1871 a mining company began work upon this coal, but abandoned it, owing to difficulties that were encountered. It is now believed that these miners did not prove the product to be of an unprofitable character, and that farther exploration is fully justified by what is known of the field. Of inferior forms of coal there is every indication of an abundance on the main-land of the province. " The tertiary or Laramie coal measures of Puget Sound and Bellingham Bay" (in the United States) " are continuous north of the international boundary, and must underlie nearly 18,000 square miles of the low country about the estuary of the Fraser and in the lower part of its valley." It is quite possible, since the better coals of Nanaimo and Comox are in demand in the San Francisco market, even at their high price and with the duty added, that these lignite fields may be worked for local consumption.

Already the value of the fish caught in the British Columbian waters is estimated at $5,000,000 a year, and yet the industry is rather at its birth than in its infancy. All the waters in and near the province fairly swarm with fish. The rivers teem with them, the straits and fiords and gulfs abound with them, the ocean beyond is freighted with an incalculable weight of living food, which must soon be distributed among the homes of the civilized world. The principal varieties of fish are the salmon, cod, shad, whitefish, bass, flounder, skate, sole, halibut, sturgeon, oolachan, herring, trout, haddock, smelts, anchovies,

THE SALMON CACHE

dog-fish, perch, sardines, oysters, crayfish shrimps, crabs, and mussels. Of other denizens of the water, the whale, sea-otter, and seal prove rich prey for those who search for them.

The main salmon rivers are the Fraser, Skeena, and Nasse rivers, but the fish also swarm in the inlets into which smaller streams empty. The Nimkish, on Vancouver Island, is also a salmon stream. Setting aside the stories of water so thick with salmon that a man might walk upon their backs, as well as that tale of the stage-coach which was upset by salmon banking themselves against it when it was crossing a fording-place, there still exist absolutely trustworthy accounts of swarms which at their height cause the largest rivers to seem alive with these fish. In such cases the ripple of their back fins frets the entire surface of the stream. I have seen photographs that show the fish in incredible numbers, side by side, like logs in a raft, and I have the word of a responsible man for the statement that he has gotten all the salmon needed for a small camp, day after day, by walking to the edge of a river and jerking the fish out with a common poker.

There are about sixteen canneries on the Fraser, six on the Skeena, three on the Nasse, and three scattered in other waters—River Inlet and Alert Bay. The total canning in 1889 was 414,294 cases, each of 48 one-pound tins. The fish are sold to Europe, Australia, and eastern Canada. The American market takes the Columbia River salmon. A round $1,000,000 is invested in the vessels, nets, trawls, canneries, oil-factories, and freezing and salting stations used in this industry in British Columbia, and about 5500 men are employed. "There is no difficulty in catching the fish," says a local historian, "for in some streams they are so crowded that they can readily be picked out of the water by hand." However, gill-nets are found to be preferable, and the fish are caught in these, which are stretched across the streams, and handled by men in flat-bottomed boats. The fish are loaded into scows and transported to the canneries, usually frame structures built upon piles close to the shores of the rivers. In the canneries the tins are made, and, as a rule, saw-mills near by produce the wood for the manufacture of the packing-cases. The fish are cleaned, rid of their heads and tails, and then chopped up and loaded into the tins by Chinamen and Indian women. The tins are then boiled, soldered, tested, packed, and shipped away. The industry is rapidly extending, and fresh salmon are now being shipped, frozen, to the markets of eastern America and England. My figures for 1889 (obtained from the Victoria *Times*) are in all likelihood under the mark for the season of 1890. The coast is made ragged by inlets, and into nearly every one a watercourse empties. All the larger streams are the haven of salmon in the spawning season, and in time the principal ones will be the bases of canning operations.

The Dominion Government has founded a salmon hatchery on the Fraser, above New Westminster. It is under the supervision of Thomas Mowat, Inspector of Fisheries, and millions of small fry are now annually turned into the great river. Whether the unexampled run of 1889 was in any part due to this process cannot be said, but certainly the salmon are not diminishing in numbers. It was feared that the refuse from the canneries would injure the "runs" of live fish, but it is now believed that there is a profit to be derived from treating the refuse for oil and guano, so that it is more likely to be saved than thrown back into the streams in the near future.

The oolachan, or candle-fish, is a valuable product of these waters, chiefly of the Fraser and Nasse rivers. They are said to be delicious when fresh, smoked, or salted, and I have it on the authority of the little pamphlet "British Columbia," handed me by a government official, that "their oil is considered superior to cod-liver oil, or any other fish-oil known." It is said that this oil is whitish, and of the consistency of thin lard. It is used as food by the natives, and is an article of barter between the coast Indians and the tribes of the interior. There is so much of it in a candle-fish of ordinary size that when one of them is dried, it will burn like a candle. It is the custom of the natives on the coast to catch the fish in immense numbers in purse-nets. They then boil them in iron-bottomed bins, straining the product in willow baskets, and running the oil into cedar boxes holding fifteen gallons each. The Nasse River candle-fish are the best. They begin running in March, and continue to come by the million for a period of several weeks.

Codfish are supposed to be very plentiful, and to frequent extensive banks at sea, but these shoals have not been explored or charted by the Government, and private enterprise will not attempt the work. Similar banks off the Alaska coast are already the resorts of California fishermen, who drive a prosperous trade in salting large catches there. The skil, or black cod, formerly known as the "coal-fish," is a splendid deep-water product. These cod weigh from eight to twenty pounds, and used to be caught by the Indians with hook and line. Already white men are driving the Indians out by superior methods. Trawls of 300 hooks are used, and the fish are found to be plentiful, especially off the west coast of the Queen Charlotte Islands. The fish is described as superior to the cod of Newfoundland in both oil and meat. The general market is not yet accustomed to it, but such a ready sale is found for what are caught that the number of vessels engaged in this fishing increases year by year. It is evident that the catch of skil will soon be an important source of revenue to the province.

AN IDEAL OF THE COAST

Herring are said to be plentiful, but no fleet is yet fitted out for them. Halibut are numerous and common. They are often of very great size. Sturgeon are found in the Fraser, whither they chase the salmon. One weighing 1400 pounds was exhibited in Victoria a few years ago, and those that weigh more than half as much are not unfrequently captured. The following is a report of the yield and value of the fisheries of the province for 1889:

Kind of Fish.		Quantity.	Value.
Salmon in cans	lbs.	20,122,128	$2,414,655 36
" fresh	lbs.	2,187,000	218,700 00
" salted	bbls.	3,749	37,460 00
" smoked	lbs.	12,900	2,580 00
Sturgeon, fresh		318,600	15,930 00
Halibut, "		605,050	30,152 50
Herring, "		190,000	9,500 00
" smoked		33,000	3,300 00
Oolachans, "		82,500	8,250 00
" fresh		6,700	1,340 00
" salted	bbls.	380	3,800 00
Trout, fresh	lbs.	14,025	1,402 50
Fish, assorted		322,725	16,136 25
Smelts, fresh		52,100	3,126 00
Rock cod		39,250	1,962 50
Skil, salted	bbls.	1,560	18,720 00
Fooshqua, fresh		268,350	13,417 50
Fur seal-skins	No.	33,570	335,700 00
Hair "	"	7,000	5,250 00
Sea-otter skins	"	115	11,500 00
Fish oil	gals.	141,420	70,710 00
Oysters	sacks	3,000	5,250 00
Clams	"	3,500	6,125 00
Mussels	"	250	500 00
Crabs	No.	175,000	5,250 00
Abelones	boxes	100	500 00
Isinglass	lbs.	5,000	1,750 00
Estimated fish consumed in province		......	100,000 00
Shrimps, prawns, etc.		......	5,000 00
Estimated consumption by Indians—			
Salmon		......	2,732,500 00
Halibut		......	190,000 00
Sturgeon and other fish		......	260,000 00
Fish oils		......	75,000 00
Approximate yield		......	$6,605,467 61

When it is considered that this is the showing of one of the newest communities on the continent, numbering only the population of what we would call a small city, suffering for want of capital and nearly all that capital brings with it, there is no longer occasion for surprise at the provincial boast that they possess far more extensive and richer fishing-fields than any on the Atlantic coast. Time and enterprise will surely test this assertion, but it is already evident that there is a vast revenue to be wrested from those waters.

I have not spoken of the sealing, which yielded $236,000 in 1887, and may yet be decided to be exclusively an American and not a British Columbian source of profit. Nor have I touched upon the extraction of oil from herrings and from dog-fish and whales, all of which are small channels of revenue.

I enjoyed the good-fortune to talk at length with a civil engineer of high repute who has explored the greater part of southern British Columbia—at least in so far as its main valleys, waterways, trails, and mountain passes are concerned. Having learned not to place too high a value upon the printed matter put forth in praise of any new country, I was especially pleased to obtain this man's practical impressions concerning the store and quality and kinds of timber the province contains. He said, not to use his own words, that timber is found all the way back from the coast to the Rockies, but it is in its most plentiful and majestic forms on the west slope of those mountains and on the west slope of the Coast Range. The very largest trees are between the Coast Range and the coast. The country between the Rocky Mountains and the Coast Range is dry by comparison with the parts where the timber thrives best, and, naturally, the forests are inferior. Between the Rockies and the Kootenay River cedar and tamaracks reach six and eight feet in diameter, and attain a height of 200 feet not infrequently. There are two or three kinds of fir and some pines (though not very many) in this region. There is very little leaf-wood, and no hard-wood. Maples are found, to be sure, but they are rather more like bushes than trees to the British Columbian mind. As one moves westward the same timber prevails, but it grows shorter and smaller until the low coast country is reached. There, as has been said, the giant forests occur again. This coast region is largely a flat country, but there are not many miles of it.

To this rule, as here laid down, there are some notable exceptions. One particular tree, called there the bull-pine—it is the pine of Lake Superior and the East—grows to great size all over the province. It is a common thing to find the trunks of these trees measuring four feet in diameter, or nearly thirteen feet in circumference. It is not especially valuable for timber, because it is too sappy. It is short-lived when exposed to the weather, and is therefore not in demand for railroad work; but for the ordinary uses to which builders put timber it answers very well.

There is a maple which attains great size at the coast, and which, when dressed, closely resembles bird's-eye-maple. It is called locally the vine-maple. The trees are found with a diameter of two-and-a-half to three feet, but the trunks seldom rise above forty or fifty feet. The wood is crooked. It runs very badly. This, of course, is what gives it the beautiful grain it possesses, and which must, sooner

THE POTLATCH

or later, find a ready market for it. There is plenty of hemlock in the province, but it is nothing like so large as that which is found in the East, and its bark is not so thick. Its size renders it serviceable for nothing larger than railway ties, and the trees grow in such inaccessible places, half-way up the mountains, that it is for the most part unprofitable to handle it. The red cedars—the wood of which is consumed in the manufacture of pencils and cigar-boxes—are also small. On the other hand, the white cedar reaches enormous sizes, up to fifteen feet of thickness at the base, very often. It is not at all extraordinary to find these cedars reaching 200 feet above the ground, and one was cut at Port Moody, in clearing the way for the railroad, that had a length of 310 feet. When fire rages in the provincial forests, the wood of these trees is what is consumed, and usually the trunks, hollow and empty, stand grimly in their places after the fire would otherwise have been forgotten. These great tubes are often of such dimensions that men put windows and doors in them and use them for dwellings. In the valleys are immense numbers of poplars of the common and cottonwood species, white birch, alder, willow, and yew trees, but they are not estimated in the forest wealth of the province, because of the expense that marketing them would entail.

This fact concerning the small timber indicates at once the primitive character of the country, and the vast wealth it possesses in what might be called heroic timber—that is, sufficiently valuable to force its way to market even from out that unopened wilderness.

It was the opinion of the engineer to whom I have referred that timber land which does not attract the second glance of a prospector in British Columbia would be considered of the first importance in Maine and New Brunswick. To put it in another way, riverside timber land which in those countries would fetch fifty dollars the acre solely for its wood, in British Columbia would not be taken up. In time it may be cut, undoubtedly it must be, when new railroads alter its value, and therefore it is impossible even roughly to estimate the value of the provincial forests.

A great business is carried on in the shipment of ninety-foot and one-hundred-foot Douglas fir sticks to the great car-building works of our country and Canada. They are used in the massive bottom frames of palace cars. The only limit that has yet been reached in this industry is not in the size of the logs, but in the capacities of the saw-mills, and in the possibilities of transportation by rail, for these logs require three cars to support their length. Except for the valleys, the whole vast country is enormously rich in this timber, the mountains (excepting the Rockies) being clothed with it from their bases to their tops. Vancouver Island is a heavily and valuably timbered country. It bears the same trees as the main-land, except that it has the oak-tree, and does not possess the tamarack. The Vancouver Island oaks do not exceed two or two-and-a-half feet in diameter. The Douglas fir (our Oregon pine) grows to tremendous proportions, especially on the north end of the island. In the old offices of the Canadian Pacific Railway at Vancouver are panels of this wood that are thirteen feet across, showing that they came from a tree whose trunk was forty feet in circumference. Tens of thousands of these firs are from eight to ten feet in diameter at the bottom.

Other trees of the province are the great silver-fir, the wood of which is not very valuable; Englemann's spruce, which is very like white spruce, and is very abundant; balsam-spruce, often exceeding two feet in diameter; the yellow or pitch pine; white pine; yellow cypress; crab-apple, occurring as a small tree or shrub; western birch, common in the Columbia region; paper or canoe birch, found sparingly on Vancouver Island and on the lower Fraser, but in abundance and of large size in the Peace River and upper Fraser regions; dogwood, arbutus, and several minor trees. Among the shrubs which grow in abundance in various districts or all over the province are the following: hazel, red elder, willow, barberry, wild red cherry, blackberry, yellow plum, choke-cherry, raspberry, gooseberry, bearberry, currant, and snowberry, mooseberry, bilberry, cranberry, whortleberry, mulberry, and blueberry.

I would have liked to write at length concerning the enterprising cities of the province, but, after all, they may be trusted to make themselves known. It is the region behind them which most interests mankind, and the Government has begun, none too promptly, a series of expeditions for exploiting it. As for the cities, the chief among them and the capital, Victoria, has an estimated population of 22,000. Its business district wears a prosperous, solid, and attractive appearance, and its detached dwellings—all of frame, and of the distinctive type which marks the houses of the California towns—are surrounded by gardens. It has a beautiful but inadequate harbor; yet in a few years it will have spread to Esquimault, now less than two miles distant. This is now the seat of a British admiralty station, and has a splendid haven, whose water is of a depth of from six to eight fathoms. At Esquimault are government offices, churches, schools, hotels, stores, a naval "canteen," and a dry-dock 450 feet long, 26 feet deep, and 65 feet wide at its entrance. The electric street railroad of Victoria was extended to Esquimault in the autumn of 1890. Of the climate of Victoria Lord Lorne said, "It is softer and more constant than that of the south of England."

Vancouver, the principal city of the main-land, is slightly smaller than Victoria, but did not begin to displace the forest until 1886. After that every house except one was destroyed by fire. To-day it boasts a hotel comparable in most important respects with any in Canada, many noble business buildings of brick or stone, good schools, fine churches, a really great area of streets built up with dwellings, and a notable system of wharves, warehouses, etc. The Canadian Pacific Railway terminates here, and so does the line of steamers for China and Japan. The city is picturesquely and healthfully situated on an arm of Burrard Inlet, has gas, water, electric lights, and shows no sign of halting its hitherto rapid growth. Of New Westminster, Nanaimo, Yale, and the still smaller towns, there is not opportunity here for more than naming.

In the original settlements in that territory a peculiar institution occasioned gala times for the red men now and then. This was the "potlatch," a thing to us so foreign, even in the impulse of which it is begotten, that we have no word or phrase to give its meaning. It is a feast and merrymaking at the expense of some man who has earned or saved what he deems considerable wealth, and who desires to distribute every iota of it at once in edibles and drinkables among the people of his tribe or village. He does this because he aspires to a chieftainship, or merely for the credit of a "potlatch"—a high distinction. Indians have been known to throw away such a sum of money that their "potlatch" has been given in a huge shed built for the feast, that hundreds have been both fed and made drunk, and that blankets and ornaments have been distributed in addition to the feast.

The custom has a new significance now. It is the white man who is to enjoy a greater than all previous potlatches in that region. The treasure has been garnered during the ages by time or nature or whatsoever you may call the host, and the province itself is offered as the feast.

IX

DAN DUNN'S OUTFIT

AT Revelstoke, 380 miles from the Pacific Ocean, in British Columbia, a small white steamboat, built on the spot, and exposing a single great paddle-wheel at her stern, was waiting to make another of her still few trips through a wilderness that, but for her presence, would be as completely primitive as almost any in North America. Her route lay down the Columbia River a distance of about one hundred and thirty miles to a point called Sproat's Landing, where some rapids interrupt navigation. The main load upon the steamer's deck was of steel rails for a railroad that was building into a new mining region in what is called the Kootenay District, just north of our Washington and Idaho. The sister range to the Rockies, called the Selkirks, was to be crossed by the new highway, which would then connect the valley of the Columbia with the Kootenay River. There was a temptation beyond the mere chance to join the first throng that pushed open a gateway and began the breaking of a trail in a brand-new country. There was to be witnessed the propulsion of civilization beyond old confines by steam-power, and this required railroad building in the Rockies, where that science finds its most formidable problems. And around and through all that was being done pressed a new population, made up of many of the elements that produced our old-time border life, and gave birth to some of the most picturesque and exciting chapters in American History.

It should be understood that here in the very heart of British Columbia only the watercourses have been travelled, and there was neither a settlement nor a house along the Columbia in that great reach of its valley between our border and the Canadian Pacific Railway, except at the landing at which this boat stopped.

Over all the varying scene, as the boat ploughed along, hung a mighty silence; for almost the only life on the deep wooded sides of the mountains was that of stealthy game. At only two points were any human beings lodged, and these were wood-choppers who supplied the fuel for the steamer—a Chinaman in one place, and two or three white men farther on. In this part of its magnificent valley the Columbia broadens in two long loops, called the Arrow Lakes, each more than two miles wide and twenty to thirty miles in length. Their prodigious towering walls are densely wooded, and in places are snow-capped in midsummer. The forest growth is primeval, and its own luxuriance crowds it beyond the edge of the grand stream in the fretwork of fallen trunks and bushes, whose roots are bedded in the soft mass of centuries of forest débris.

Early in the journey the clerk of the steamer told me that wild animals were frequently seen crossing the river ahead of the vessel; bear, he said, and deer and elk and porcupine. When I left him to go to my state-room and dress for the rough journey ahead of me, he came to my door, calling in excited tones for me to come out on the deck. "There's a big bear ahead!" he cried, and as he spoke I saw the black head of the animal cleaving the quiet water close to the nearer shore. Presently Bruin's feet touched the bottom, and he bounded into the bush and disappeared.

The scenery was superb all the day, but at sundown nature began to revel in a series of the most splendid and spectacular effects. For an hour a haze had clothed the more distant mountains as with a transparent veil, rendering the view dream-like and soft beyond description. But as the sun sank to the summit of the uplifted horizon it began to lavish the most intense colors upon all the objects in view. The snowy peaks turned to gaudy prisms as of crystal, the wooded summits became impurpled, the nearer hills turned a deep green, and the tranquil lake assumed a bright pea-color. Above all else, the sky was gorgeous. Around its western edge it took on a rose-red blush that blended at the zenith with a deep blue, in which were floating little clouds of amber and of flame-lit pearl.

A moonless night soon closed around the boat, and in the morning we were at Sproat's Landing, a place two months old. The village consisted of a tiny cluster of frame-houses and tents perched on the edge of the steep bank of the Columbia. One building was the office and storehouse of the projected railroad, two others were general trading stores, one was the hotel, and the other habitations were mainly tents.

I firmly believe there never was a hotel like the

hostlery there. In a general way its design was an adaptation of the plan of a hen-coop. Possibly a box made of gridirons suggests more clearly the principle of its construction. It was two stories high, and contained about a baker's dozen of rooms, the main one being the bar-room, of course. After the framework

AN INDIAN CANOE ON THE COLUMBIA

had been finished, there was perhaps half enough "slab" lumber to sheathe the outside of the house, and this had been made to serve for exterior and interior walls, and the floors and ceilings besides. The consequence was that a flock of gigantic canaries might have been kept in it with propriety, but as a place of abode for human beings it compared closely with the Brooklyn Bridge.

They have in our West many very frail hotels that the people call "telephone houses," because a tenant can hear in every room whatever is spoken in any part of the building; but in this house one could stand in any room and see into all the others. A clergyman and his wife stopped in it on the night before I arrived, and the good woman stayed up until nearly daylight, pinning papers on the walls and laying them on the floor until she covered a corner in which to prepare for bed.

I hired a room and stored my traps in it, but I slept in one of the engineers' tents, and met with a very comical adventure. The tent contained two cots, and a bench on which the engineer, who occupied one of the beds, had heaped his clothing. Supposing him to be asleep, I undressed quietly, blew out the candle, and popped into my bed. As I did so one pair of its legs broke down, and it naturally occurred to me, at almost the same instant, that the bench was of about the proper height to raise the fallen end of the cot to the right level.

"Broke down, eh?" said my companion—a man, by-the-way, whose face I have never yet seen.

"Yes," I replied. "Can I put your clothing on the floor and make use of that bench?"

"Aye, that you can."

So out of bed I leaped, put his apparel in a heap on the floor, and ran the bench under my bed. It proved to be a neat substitute for the broken legs, and I was quickly under the covers again and ready for sleep.

The engineer's voice roused me.

"That's what I call the beauty of a head-piece," he said. Presently he repeated the remark.

"Are you speaking to me?" I asked.

"Yes; I'm saying that's what I call the beauty of a head-piece. It's wonderful; and many's the day and night I'll think of it, if I live. What do I mean? Why, I mean that that is what makes you Americans such a great people—it's the beauty of having head-pieces on your shoulders. It's so easy to think quick if you've got something to think with. Here you are, and your bed breaks down. What would I do? Probably nothing. I'd think what a beastly scrape it was, and I'd keep on thinking till I went to sleep. What do you do? Why, as quick as a flash you says, 'Hello, here's a go!' 'May I have the bench?' says you. 'Yes,' says I. Out of bed you go, and you clap the bench under the bed, and there you are, as right as a trivet. That's the beauty of a head-piece, and that's what makes America the wonderful country she is."

Never was a more sincere compliment paid to my country, and I am glad I obtained it so easily.

There was a barber pole in front of the house, set up by a "prospector" who had run out of funds (and everything else except hope), and who, like all his kind, had stopped to "make a few dollars" wherewith to outfit again and continue his search for gold. He noted the local need of a barber, and instantly became one by purchasing a razor on credit, and painting a pole while waiting for custom. He was a jocular fellow—a born New Yorker, by-the-way.

"Don't shave me close," said I.

"Close?" he repeated. "You'll be the luckiest victim I've slashed yet if I get off any of your beard at all. How's the razor?"

"All right."

"Oh no, it ain't," said he; "you're setting your nerves to stand it, so's not to be called a tender-foot. I'm no barber. I expected to 'tend bar when I bumped up agin this place. If you could see the blood streaming down your face you'd faint."

In spite of his self-depreciation, he performed as artistic and painless an operation as I ever sat through.

While I was being shaved the loungers in the barber-shop entered into a conversation that revealed, as nothing else could have disclosed it, the deadly monotony of life in that little town. A hen cackled out-of-doors, and the loungers fell to questioning one another as to which hen had laid an egg.

"It must be the black one," said the barber.

"Yet it don't exactly sound like old blacky's cackle," said a more deliberate and careful speaker.

"'Pears to me 's though it might be the speckled un," ventured a third.

"She ain't never laid no eggs," said the barber.

"Could it be the bantam?" another inquired.

Thus they discussed with earnestness this most interesting event of the morning, until a young man darted into the room with his eyes lighted by excitement.

"Say, Bill," said he, almost breathlessly, "that's the speckled hen a-cackling, by thunder! She's laid an egg, I guess."

In Sproat's Landing we saw the nucleus of a railroad terminal point. The queer hotel was but little more peculiar than many of the people who gathered

"YOU'RE SETTING YOUR NERVES TO STAND IT"

on the single street on pay-day to spend their hard-earned money upon a great deal of illicit whiskey and a few rude necessaries from the limited stock on sale in the stores. There never had been any grave disorder there, yet the floating population was as motley a collection of the riffraff of the border as one could well imagine, and there was only one policeman to enforce the law in a territory the size of Rhode Island. He was quite as remarkable in his way as any other development of that embryotic civilization. His name was Jack Kirkup, and all who knew him spoke of him as being physically the most superb example of manhood in the Dominion. Six feet and three inches in height, with the chest, neck, and limbs of a giant, his three hundred pounds of weight were so exactly his complement as to give him the symmetry of an Apollo. He was good-looking, with the beauty of a round-faced, good-natured boy, and his thick hair fell in a cluster of ringlets over his forehead and upon his neck. No knight of Arthur's circle can have been more picturesque a figure in the forest than this "Jack." He was as neat as a dandy. He wore high boots and corduroy knickerbockers, a flannel shirt and a sack-coat, and rode his big bay horse with the ease and grace of a Skobeleff. He smoked like a fire of green brush, but had never tasted liquor in his life. In a dozen years he had slept more frequently in the open air, upon pebble beds or in trenches in the snow, than upon ordinary bedding, and he exhibited, in his graceful movements, his sparkling eyes and ruddy cheeks, his massive frame and his imperturbable good-nature, a degree of health and vigor that would seem insolent to the average New Yorker. Now that the railroad was building, he kept ever on the trail, along what was called "the right of way"—going from camp to camp to "jump" whiskey peddlers and gamblers and to quell disorder—except on pay-day, once a month, when he stayed at Sproat's Landing.

The echoes of his fearless behavior and lively adventures rang in every gathering. The general tenor of the stories was to the effect that he usually gave one warning to evil-doers, and if they did not heed that he "cleaned them out." He carried a revolver, but never had used it. Even when the most notorious gambler on our border had crossed over into "Jack's" bailiwick the policeman depended upon his fists. He had met the gambler and had "advised" him to take the cars next day. The gambler, in re-

JACK KIRKUP, THE MOUNTAIN SHERIFF

ply, had suggested that both would get along more quietly if each minded his own affairs, whereupon Kirkup had said, "You hear me: take the cars out of here to-morrow." The little community (it was Donald, B. C., a very rough place at the time) held its breathing for twenty-four hours, and at the approach of train-time was on tiptoe with strained anxiety. At twenty minutes before the hour the policeman, amiable and easy-going as ever in appearance, began a tour of the houses. It was in a tavern that he found the gambler.

"You must take the train," said he.

"You can't make me," replied the gambler.

There were no more words. In two minutes the giant was carrying the limp body of the ruffian to a wagon, in which he drove him to the jail. There he washed the blood off the gambler's face and tidied his collar and scarf. From there the couple walked to the cars, where they parted amicably.

"I had to be a little rough," said Kirkup to the loungers at the station, "because he was armed like a pin-cushion, and I didn't want to have to kill him."

We made the journey from Sproat's Landing to the Kootenay River upon a sorry quartet of packhorses that were at other times employed to carry provisions and material to the construction camps. They were of the kind of horses known all over the West as "cayuses." The word is the name of a once notable tribe of Indians in what is now the State of Washington. To these Indians is credited the introduction of this small and peculiar breed of horses, but many persons in the West think the horses get the nickname because of a humorous fancy begotten of their wildness, and suggesting that they are only part horses and part coyotes. But all the wildness and the characteristic "bucking" had long since been "packed" out of these poor creatures, and they needed the whip frequently to urge them upon a slow progress. Kirkup was going his rounds, and accompanied us on our journey of less than twenty miles to the Kootenay River. On the way one saw every stage in the construction of a railway. The process of development was reversed as we travelled, because the work had been pushed well along where we started, and was but at its commencement where we ended our trip. At the landing half a mile or more of the railroad had been completed, even to the addition of a locomotive and two gondola cars. Beyond the little strip of rails was a long reach of graded roadbed, and so the progress of the work dwindled, until at last there was little more than the trail-cutters' path to mark what had been determined as the "right of way."

For the sake of clearness, I will first explain the steps that are taken at the outset in building a railroad, rather than tell what parts of the undertaking we came upon in passing over the various "contracts" that were being worked in what appeared a confusing and hap-hazard disorder. I have mentioned that one of the houses at the landing was the railroad company's storehouse, and that near by were the tents of the surveyors or civil engineers. The road was to be a branch of the Canadian Pacific system, and these engineers were the first men sent into the country, with instructions to survey a line to the new mining region, into which men were pouring from the older parts of Canada and from our country. It was understood by them that they were to hit upon the most direct and at the same time the least expensive route for the railroad to take. They went to the scene of their labors by canoes, and carried tents, blankets, instruments, and what they called their "grub stakes," which is to say, their food. Then they travelled over the ground between their two terminal points, and back by another route, and back again by still another route, and so back and forth perhaps four and possibly six times. In that way alone were they enabled to select the line which offered the shortest length and the least obstacles in number and degree for the workmen who were to come after them.

At Sproat's Landing I met an engineer, Mr. B. C. Stewart, who is famous in his profession as the most tireless and intrepid exponent of its difficulties in the Dominion. The young men account it a misfortune to be detailed to go on one of his journeys with him. It is his custom to start out with a blanket, some bacon and meal, and a coffee-pot, and to be gone for weeks, and even for months. There scarcely can have been a hardier Scotchman, one of more simple tastes and requirements, or one possessing in any higher degree the quality called endurance. He has spent years in the mountains of British Columbia, finding and exploring the various passes, the most direct and feasible routes to and from them, the valleys between the ranges, and the characteristics of each section of

ENGINEER ON THE PRELIMINARY SURVEY

the country. In a vast country that has not otherwise been one-third explored he has made himself familiar with the full southern half. He has not known what it was to enjoy a home, nor has he seen an apple growing upon a tree in many years. During his long and close-succeeding trips he has run the whole gamut of the adventures incident to the lives of hunters or explorers, suffering hunger, exposure, peril from wild beasts, and all the hair-breadth escapes from frost and storm and flood that Nature unvanquished visits upon those who first brave her depths. Such is the work and such are the men that figure in the foremost preliminaries to railroad building.

Whoever has left the beaten path of travel or gone beyond a well-settled region can form a more or less just estimate of that which one of these professional pioneers encounters in prospecting for a railroad. I had several "tastes," as the Irish express it, of that very Kootenay Valley. I can say conscientiously that I never was in a wilder region. In going only a few yards from the railroad "right of way" the difficulties of an experienced pedestrianism like my own instantly became tremendous. There was a particularly choice spot for fishing at a distance of three-quarters of a mile from Dan Dunn's outfit, and I travelled the road to it half a dozen times. Bunyan would have strengthened the *Pilgrim's Progress* had he known of such conditions with which to surround his hero. Between rocks the size of a city mansion and unsteady bowlders no larger than a man's head the ground was all but covered. Among this wreckage trees grew in wild abundance, and countless trunks of dead ones lay rotting between them. A jungle as dense as any I ever saw was formed of soft-wood saplings and bushes, so that it was next to impossible to move a yard in any direction. It was out of the question for any one to see three yards ahead, and there was often no telling when a foot was put down whether it was going through a rotten trunk or upon a spinning bowlder, or whether the black shadows here and there were a foot deep or were the mouths of fissures that reached to China. I fished too long one night, and was obliged to make that journey after dark. After ten minutes crowded with falls and false steps, the task seemed so hopelessly impossible that I could easily have been induced to turn back and risk a night on the rocks at the edge of the tide.

It was after a thorough knowledge of the natural conditions which the railroad men were overcoming that the gradual steps of their progress became most interesting. The first men to follow the engineers, after the specifications have been drawn up and the contracts signed, are the "right-of-way" men. These are partly trail-makers and partly laborers at the heavier work of actually clearing the wilderness for the road-bed. The trail-cutters are guided by the long line of stakes with which the engineers have marked the course the road is to take. The trail-men are sent out to cut what in general parlance would be called a path, over which supplies are to be thereafter carried to the workmen's camps. The path they cut must therefore be sufficiently wide for the passage along it of a mule and his load. As a mule's load will sometimes consist of the framework of a kitchen range, or the end boards of a bedstead, a five-foot swath through the forest is a trail of serviceable width. The trail-cutters fell the trees to right and left, and drag the fallen trunks out of the path as they go along, travelling and working between a mile and two miles each day, and moving their tents and provisions on pack-horses as they advance. They keep reasonably close to the projected line of the railway, but the path they cut is apt to be a winding one that avoids the larger rocks and the smaller ravines. Great distortions, such as hills or gullies, which the railroad must pass through or over, the trail men pay no heed to; neither do the pack-horses, whose tastes are not consulted, and who can cling to a rock at almost any angle, like flies of larger growth. This trail, when finished, leads from the company's storehouse all along the line, and from that storehouse, on the backs of the pack-animals, come all the food and tools and clothing, powder, dynamite, tents, and living utensils, to be used by the workmen, their bosses, and the engineers.

Slowly, behind the trail-cutters, follow the "right-of-way" men. These are axemen also. All that they do is to cut the trees down and drag them out of the way.

It is when the axemen have cleared the right of way that the first view of the railroad in embryo is obtainable. And very queer it looks. It is a wide avenue through the forest, to be sure, yet it is little like any forest drive that we are accustomed to in the realms of civilization.

Every succeeding stage of the work leads towards

the production of an even and level thoroughfare, without protuberance or depression, and in the course of our ride to Dan Dunn's camp on the Kootenay we saw the rapidly developing railroad in each phase of its evolution from the rough surface of the wilderness. Now we would come upon a long reach of finished road-bed on comparatively level ground all ready for the rails, with carpenters at work in little gullies which they were spanning with timber trestles. Next we would see a battalion of men and dump-carts cutting into a hill of dirt and carting its substance to a neighboring valley, wherein they were slowly heaping a long and symmetrical wall of earth-work, with sloping sides and level top, to bridge the gap between hill and hill.

FALLING MONARCHS

Again, we came upon places where men ran towards us shouting that a "blast" was to be fired. Here was what was called "rockwork," where some granite rib of a mountain or huge rocky knoll was being blown to flinders with dynamite.

And so, through all these scenes upon the pack-trail, we came at last to a white camp of tents hidden in the lush greenery of a luxuriant forest, and nestling beside a rushing mountain torrent of green water flecked with the foam from an eternal battle with a myriad of sunken rocks. It was Dunn's headquarters—the construction camp. Evening was falling, and the men were clambering down the hill-side trails from their work. There was no order in the disposition of the tents, nor had the forest been prepared for them. Their white sides rose here and there wherever there was a space between the trees, as if so many great white moths had settled in a garden. Huge trees had been felled and thrown across ravines to serve as aerial foot-paths from point to point, and at the river's edge two or three tents seemed to have been pushed over the steep bluff to find lodgement on the sandy beach beside the turbulent stream.

There were other camps on the line of this work, and it is worth while to add a word about their management and the system under which they were maintained. In the first place, each camp is apt to be the outfit of a contractor. The whole work of building a railroad is let out in contracts for portions of five, ten, or fifteen miles. Even when great jobs of seventy or a hundred miles are contracted for in one piece, it is customary for the contractor to divide his task and sublet it. But a fairly representative bit of mountain work is that which I found Dan Dunn superintending, as the factotum of the contractor who undertook it.

If a contractor acts as "boss" himself, he stays upon the ground; but in this case the contractor had other undertakings in hand. Hence the presence of Dan Dunn, his walking boss or general foreman. Dunn is a man of means, and is himself a contractor by profession, who has worked his way up from a start as a laborer.

The camp to which we came was a portable city, complete except for its lack of women. It had its artisans, its professional men, its store and workshops, its seat of government and officers, and its policeman, its amusement hall, its work-a-day and social sides. Its main peculiarity was that its boss (for it was like an American city in the possession of that functionary also) had announced that he was going to move it a couple of miles away on the following Sunday. One tent was the stableman's, with a capacious "corral" fenced in near by for the keeping of the pack horses and mules. His corps of assistants was a large one; for, besides the pack-horses that connected the camp with the outer world, he had the keeping of all the "grade-horses," so called—those which draw the stone and dirt carts and the little dump-cars on the false tracks set up on the levels near where "filling" or "cutting" is to be done. Another tent was the blacksmith's. He had a "helper," and was a busy man, charged with all the tool-sharpening, the care of all the horses' feet, and the repairing of all the ironwork of the wagons, cars, and dirt-scrapers. Near by was the harness-man's tent, the shop of the leather-mender. In the centre of the camp, like a low citadel, rose a mound of logs and earth bearing on a sign the single word "Powder," but containing within its great sunken chamber a considerable store of various explosives — giant, black, and Judson powder, and dynamite.

More tremendous force is used in railroad blasting than most persons imagine. In order to perform a quick job of removing a section of solid mountain, the drill-men, after making a bore, say, twenty feet in depth, begin what they call "springing" it by explod-

DAN DUNN ON HIS WORKS

ing little cartridges in the bottom of the drill hole until they have produced a considerable chamber there. The average amount of explosive for which they thus prepare a place is 40 or 50 kegs of giant powder and 10 kegs of black powder; but Dunn told me he had seen 280 kegs of black powder and

500 pounds of dynamite used in a single blast in mountain work.

Another tent was that of the time-keeper. He journeyed twice a day all over the work, five miles up and five down. On one journey he noted what men were at labor in the forenoon, and on his return he tallied those who were entitled to pay for the second half of the day. Such an official knows the name of every laborer, and, moreover, he knows the pecuniary rating of each man, so that when the workmen stop him to order shoes or trousers, blankets, shirts, tobacco, penknives, or what not, he decides upon his own responsibility whether they have sufficient money coming to them to meet the accommodation.

The "store" was simply another tent. In it was kept a fair supply of the articles in constant demand — a supply brought from the headquarters store at the other end of the trail, and constantly replenished by the pack-horses. This trading-place was in charge of a man called "the book-keeper," and he had two or three clerks to assist him. The stock was precisely like that of a cross-roads country store in one of our older States. Its goods included simple medicines, boots, shoes, clothing, cutlery, tobacco, cigars, pipes, hats and caps, blankets, thread and needles, and several hundred others among the ten thousand necessaries of a modern laborer's life. The only legal tender received there took the shape of orders written by the time-keeper, for the man in charge of the store was not required to know the ratings of the men upon the pay-roll.

The doctor's tent was among the rest, but his office

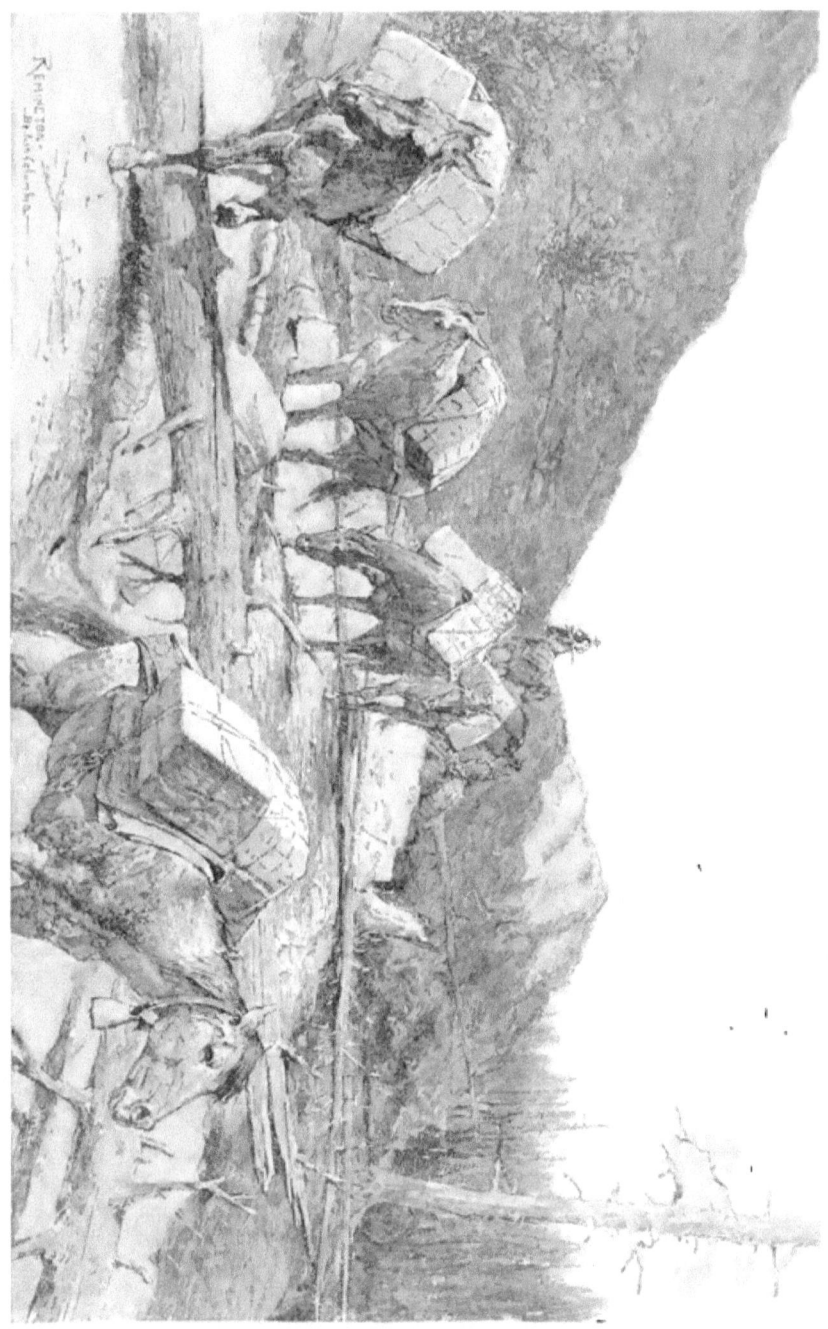

might aptly have been said to be "in the saddle." He was nominally employed by the company, but each man was "docked," or charged, seventy-five cents a month for medical services whether he ever needed a doctor or not. When I was in the camp there was only one sick man—a rheumatic. He had a tent all to himself, and his meals were regularly carried to him. Though he was a stranger to every man there, and had worked only one day before he surrendered to sickness, a purse of about forty dollars had been raised for him among the men, and he was to be "packed" to Sproat's Landing on a mule at the company's expense whenever the doctor decreed it wise to move him. Of course invalidism of a more serious nature is not infrequent where men work in the paths of sliding rocks, beneath caving earth, amid falling forest trees, around giant blasts, and with heavy tools.

Another one of the tents was that of the "boss packer." He superintended the transportation of supplies on the pack-trail. This "job of 200 men," as Dunn styled his camp, employed thirty pack horses and mules. The pack-trains consisted of a "bell-horse" and boy, and six horses following. Each animal was rated to carry a burden of 400 pounds of dead weight, and to require three quarts of meal three times a day.

Another official habitation was the "store-man's" tent. As a rule, there is a store-man to every ten miles of construction work; often every camp has one. The store-man keeps account of the distribution of the supplies of food. He issues requisitions upon the head storehouse of the company, and makes out orders for each day's rations from the camp store. The cooks are therefore under him, and this fact suggests a mention of the principal building in the camp —the mess hall, or "grub tent."

This structure was of a size to accommodate two hundred men at once. Two tables ran the length of the unbroken interior—tables made roughly of the slabs or outside boards from a saw-mill. The benches were huge tree-trunks spiked fast upon stumps. There was a bench on either side of each table, and the places for the men were each set with a tin cup and a tin pie plate. The bread was heaped high on wooden platters, and all the condiments — catsup, vinegar, mustard, pepper, and salt—were in cans that had once held condensed milk. The cooks worked in an open-ended extension at the rear of the great room. The rule is to have one cook and two "cookees" to each sixty men.

While I was a new arrival just undergoing introduction, the men, who had come in from work, and who had "washed up" in the little creeks and at the river bank, began to assemble in the "grub tent" for supper. They were especially interesting to me because there was every reason to believe that they formed an assembly as typical of the human flotsam of the border as ever was gathered on the continent. Very few were what might be called born laborers; on the contrary, they were mainly men of higher origin who had failed in older civilizations; outlaws from the States; men who had hoped for a gold-mine until hope was all but dead; men in the first flush of the gold fever; ne'er-do-wells; and here and there a working-man by training. They ate as a good many other sorts of men do, with great rapidity, little etiquette, and just enough unselfishness to pass each other the bread. It was noticeable that they seemed to have no time for talking. Certainly they had earned the right to be hungry, and the food was good and plentiful.

A SKETCH ON THE WORK

Dan Dunn's tent was just in front of the mess tent, a few feet away on the edge of the river bluff. It was a little "A" tent, with a single cot on one side, a wooden chest on the other, and a small table between the two at the farther end, opposite the door.

"Are ye looking at my wolverenes?" said he. "There's good men among them, and some that ain't so good, and many that's worse. But railroading is good enough for most of 'em. It ain't too rich for any man's blood, I assure ye."

Over six feet in height, broad-chested, athletic, and carrying not an ounce of flesh that could be spared, Dan Dunn's was a striking figure even where physical strength was the most serviceable possession of every man. From never having given his personal appearance a thought—except during a brief period of courtship antecedent to the establishment of a home in old Ontario—he had so accustomed himself to unrestraint that his habitual attitude was that of a long-bladed jack-knife not fully opened. His long spare arms swung limberly before a long spare body set upon long spare legs. His costume was one that is never described in the advertisements of city clothiers. It consisted of a dust-coated slouch felt hat, which a dealer once sold for black, of a flannel shirt, of homespun trousers, of socks, and of heavy "brogans." In all, his dress was what the æsthetes of Mr. Wilde's day might have aptly termed a symphony in dust. His shoes and hat had acquired a mud-color, and his shirt and trousers were chosen because they originally possessed it. Yet Dan Dunn was distinctly a cleanly man, fond of frequent splashing in the camp toilet basins—the Kootenay River and its little rushing tributaries. He was not shaven. As a rule he is not, and yet at times he is, as it happens. I learned that on Sundays, when there was nothing to do except to go fishing, or to walk over to the engineer's camp for intellectual society, he felt the unconscious impulse of a forgotten training, and put on a coat. He even tied a black silk ribbon under his collar on such occasions, and if no one had given him a good cigar during the week, he took out his best pipe (which had been locked up, because whatever was not under lock and key was certain to be stolen in half an hour). Then he felt fitted, as he would say, "for a hard day's work at loafing."

If you came upon Dan Dunn on Broadway, he would look as awkward as any other animal removed from its element; yet on a forest trail not even Davy Crockett was handsomer or more picturesque. His face is reddish-brown and as hard-skinned as the top of a drum, befitting a man who has lived out-of-doors all his life. But it is a finely moulded face, instinct with good-nature and some gentleness. The witchery of quick Irish humor lurks often in his eyes, but

THE MESS TENT AT NIGHT

can quickly give place on occasion to a firm light, which is best read in connection with the broad, strong sweep of his massive under-jaw. There you see his fitness to command small armies, even of what he calls "wolverenes." He is willing to thrash any man who seems to need the operation, and yet he is equally noted for gathering a squad of rough laborers in every camp to make them his wards. He collects the money such men earn, and puts it in bank, or sends it to their families.

"It does them as much good to let me take it as to chuck it over a gin-mill bar," he explained.

As we stood looking into the crowded booth, where the men sat elbow to elbow, and all the knife blades were plying to and from all the plates and mouths, Dunn explained that his men were well fed.

"The time has gone by," said he, "when you could keep an outfit on salt pork and bacon. It's as far gone as them days when they say the Hudson Bay Company fed its laborers on rabbit tracks and a stick. Did ye never hear of that? Why, sure, man, 'twas only fifty years ago that when meal hours came the bosses of the big trading company would give a workman a stick, and point out some rabbit tracks, and tell him he'd have an hour to catch his fill. But in railroading nowadays we give them the best that's going, and all they want of it—beef, ham, bacon, potatoes, mush, beans, oatmeal, the choicest fish, and game right out of the woods, and every sort of vegetable (canned, of course). Oh, they must be fed well, or they wouldn't stay."

He said that the supplies of food are calculated on

the basis of three-and-a-half pounds of provisions to a man—all the varieties of food being proportioned so that the total weight will be three-and-a-half pounds a day. The orders are given frequently and for small amounts, so as to economize in the number of horses required on the pack-trail. The amount to be consumed by the horses is, of course, included in the loads. The cost of "packing" food over long distances is more considerable than would be supposed. It was estimated that at Dunn's camp the freighting cost forty dollars a ton, but I heard of places farther in the mountains where the cost was double that. Indeed, a discussion of the subject brought to light the fact that in remote mining camps the cost of "packing" brought lager-beer in bottles up to the price of champagne. At one camp on the Kootenay bacon was selling at the time I was in the valley at thirty cents a pound, and dried peaches fetched forty cents under competition.

As we looked on, the men were eating fresh beef and vegetables, with tea and coffee and pie. The head cook was a man trained in a lumber camp, and therefore ranked high in the scale of his profession. Every sort of cook drifts into camps like these, and that camp considers itself the most fortunate which happens to eat under the ministrations of a man who has cooked on a steamboat; but a cook from a lumber camp is rated almost as proudly.

"Ye would not think it," said Dunn, "but some of them men has been bank clerks, and there's doctors and teachers among 'em—everything, in fact, except preachers. I never knew a preacher to get into a

"THEY GAINED ERECTNESS BY SLOW JOLTS"

railroad gang. The men are always changing — coming and going. We don't have to advertise for new hands. The woods is full of men out of a job, and out of everything—pockets, elbows, and all. They drift in like peddlers on a pay-day. They come here with no more clothing than will wad a gun. The most of them will get nothing after two months' work. You see, they're mortgaged with their fares against them (thirty to forty dollars for them which the railroad brings from the East), and then they have their meals to pay for, at five dollars a week while they're here, and on top of that is all the clothing and shoes and blankets and tobacco, and everything they need—all charged agin them. It's just as well for them, for the most of them are too rich if they're a dollar ahead. There's few of them can stand the luxury of thirty dollars. When they get a stake of them dimensions, the most of them will stay no longer after pay-day than John Brown stayed in heaven. The most of them bang it all away for drink, and they are sure to come back again, but the 'prospectors' and chronic tramps only work to get clothes and a flirting acquaintance with food, as well as money enough to make an affidavit to, and they never come back again at all. Out of 8500 men we had in one big work in Canada, 1500 to 2000 knocked off every month. Ninety per cent. came back. They had just been away for an old-fashioned drunk."

It would be difficult to draw a parallel between these laborers and any class or condition of men in the East. They were of every nationality where news of gold-mines, of free settlers' sections, or of quick fortunes in the New World had penetrated. I recognized Greeks, Finns, Hungarians, Danes, Scotch, English, Irish, and Italians among them. Not a man exhibited a coat, and all were tanned brown, and were as spare and slender as excessively hard work can make a man. There was not a superfluity or an ornament in sight as they walked past me; not a necktie, a finger-ring, nor a watch-chain. There were some very intelligent faces and one or two fine ones in the band. Two typical old-fashioned prospectors especially attracted me. They were evidently of gentle birth, but time and exposure had bent them, and silvered their long, unkempt locks. Worse than all, it had planted in their faces a blended expression of sadness and hope fatigued that was painful to see. It is the brand that is on every old prospector's face. A very few of the men were young fellows of thirty, or even within the twenties. Their youth impelled them to break away from the table earlier than the others, and, seizing their rods, to start off for the fishing in the river.

But those who thought of active pleasure were few indeed. Theirs was killing work, the most severe kind, and performed under the broiling sun, that at high mountain altitudes sends the mercury above 100° on every summer's day, and makes itself felt as if the rarefied atmosphere was no atmosphere at all. After a long day at the drill or the pick or shovel in such a climate, it was only natural that the men should, with a common impulse, seek first the solace of their pipes, and then of the shake-downs in their tents. I did not know until the next morning how severely their systems were strained; but it happened at sunrise on that day that I was at my ablutions on the edge of the river when Dan Dunn's gong turned the silent forest into a bedlam. It was called the seven-o'clock alarum, and was rung two hours earlier than that hour, so that the men might take two hours after dinner out of the heat of the day, "else the sun would kill them," Dunn said. This was apparently his device, and he kept up the transparent deception by having every clock and watch in the camp set two hours out of time.

With the sounding of the gong the men began to appear outside the little tents in which they slept in

couples. They came stumbling down the bluff to
wash in the river, and of all the pitiful sights I ever
saw, they presented one of the worst; of all the
straining and racking and exhaustion that ever hard
labor gave to men, they exhibited the utmost. They
were but half awakened, and they moved so painfully
and stiffly that I imagined I could hear their bones
creak. I have seen spavined work-horses turned out
to die that moved precisely as these men did. It was
shocking to see them hobble over the rough ground;
it was pitiful to watch them as they attempted to
straighten their stiffened bodies after they had been
bent double over the water. They gained erectness
by slow jolts, as if their joints were of iron that had
rusted. Of course they soon regained whatever elasticity nature had left them, and were themselves for
the day—an active, muscular force of men. But that
early morning sight of them was not such a spectacle
as a right-minded man enjoys seeing his fellows take
part in.

THE END

Interesting Works
of
Travel and Exploration.

Allen's Blue-Grass Region.
The Blue-Grass Region of Kentucky, and other Kentucky Articles. By JAMES LANE ALLEN. Illustrated. 8vo, Cloth, Ornamental, $2 50.

Miss Edwards's Egypt.
Pharaohs, Fellahs, and Explorers. By AMELIA B. EDWARDS. Profusely Illustrated. 8vo, Cloth, $4 00.

Hearn's West Indies.
Two Years in the French West Indies. By LAFCADIO HEARN. Illustrated. Post 8vo, Cloth, Ornamental, $2 00.

Miss Scidmore's Japan.
Jinrikisha Days in Japan. By ELIZA RUHAMAH SCIDMORE. Illustrated. Post 8vo, Cloth, Ornamental, $2 00.

Child's South America.
Spanish-American Republics. By THEODORE CHILD. Profusely Illustrated. Square 8vo, Cloth, $3 50.

The Tsar and His People.
The Tsar and His People; or, Social Life in Russia. By THEODORE CHILD, and Others. Profusely Illustrated. Square 8vo, Cloth, Uncut Edges and Gilt Top. $3 00.

Child's Summer Holidays.
Summer Holidays. Travelling Notes in Europe. By THEODORE CHILD. Post 8vo, Cloth. $1 25.

Warner's Southern California.
Our Italy. An Exposition of the Climate and Resources of Southern California. By CHARLES DUDLEY WARNER. Illustrated. 8vo, Cloth, Ornamental, $2 50.

Warner's South and West.
Studies in the South and West, with Comments on Canada. By CHARLES DUDLEY WARNER. Post 8vo, Half Leather, $1 75.

Interesting Works of Travel and Exploration.

Curtis's Spanish America.
The Capitals of Spanish America. By WILLIAM ELEROY CURTIS. With a Colored Map and 358 Illustrations. 8vo, Cloth, Extra, $3 50.

Bridgman's Algeria.
Winters in Algeria. Written and Illustrated by FREDERICK ARTHUR BRIDGMAN. Square 8vo, Cloth, Ornamental, $2 50.

Pennells' Hebrides.
Our Journey to the Hebrides. By JOSEPH PENNELL and ELIZABETH ROBINS PENNELL. Illustrated. Post 8vo, Cloth, Ornamental, $1 75.

Miss Bisland's Trip Around the World.
A Flying Trip Around the World. By ELIZABETH BISLAND. With Portrait. 16mo, Cloth, Ornamental, $1 25.

Mrs. Custer's Two Volumes.
BOOTS AND SADDLES; or, Life in Dakota with General Custer. With Portrait.—FOLLOWING THE GUIDON. Illustrated.—By Mrs. ELIZABETH B. CUSTER. Post 8vo, Cloth, $1 50 each.

Captain King's Campaigning with Crook.
Campaigning with Crook, and Stories of Army Life. By Captain CHARLES KING, U.S.A. Illustrated. Post 8vo, Cloth, $1 25.

Mrs. Wallace's Travel Sketches.
The Storied Sea. By SUSAN E. WALLACE. 18mo, Cloth, $1 00.

Meriwether's A Tramp Trip.
A Tramp Trip. How to See Europe on Fifty Cents a Day. By LEE MERIWETHER. With Portrait. 12mo, Cloth, Ornamental, $1 25.

Nordhoff's California.
Peninsular California. Some Account of the Climate, Soil, Productions, and Present Condition chiefly of the Northern Half of Lower California. By CHARLES NORDHOFF. Maps and Illustrations. Square 8vo, Cloth, $1 00; Paper, 75 cents.

PUBLISHED BY HARPER & BROTHERS, NEW YORK.

☞ HARPER & BROTHERS *will send any of the above works by mail, postage prepaid, to any part of the United States, Canada, or Mexico, on receipt of the price.*

www.ingramcontent.com/pod-product-compliance
Lightning Source LLC
Chambersburg PA
CBHW021151230426
43667CB00006B/346